The Fundamentals and Challenges of Oxide Materials

The Fundamentals and Challenges of Oxide Materials

Edited by

Mihail Lungu and Catalin Nicolae Marin

Cambridge Scholars Publishing

The Fundamentals and Challenges of Oxide Materials

Edited by Mihail Lungu and Catalin Nicolae Marin

This book first published 2023

Cambridge Scholars Publishing

Lady Stephenson Library, Newcastle upon Tyne, NE6 2PA, UK

British Library Cataloguing in Publication Data
A catalogue record for this book is available from the British Library

ISBN (10): 1-5275-9166-2
ISBN (13): 978-1-5275-9166-0

TABLE OF CONTENTS

CHAPTER 1

METHODS OF SYNTHESIS OF OXIDES

ANTOANETTA LUNGU, PAULINA VLAZAN,
PAULA SFIRLOAGA, MARIA POIENAR

1.1. Introduction

This chapter covers the methods of synthesis of the spinel, perovskite, and crednerite-type oxides. Metal oxides are one of the best studied categories of materials, as they present a wide range of structures and properties. The method used to obtain spinelic oxides ($MnFe_2O_4$) under conditions of maximum purity and optimum efficiency is determined by the fact that the structure and properties of mixed oxide systems depend primarily on the method. As such, five categories of methods are presented. These are chemical co-precipitation, direct calcination, and the hydrothermal method (including the classical hydrothermal method, and the ultrasound and microwave-assisted hydrothermal methods). Some of these methods, along with others (direct reaction in sealed silica tubes or ultrasound assisted co-preciptation, for example, in the case of crednerite $CuMnO_2$ materials) have been used to obtain crednerite materials and will be described in the present book. To obtain perovskite materials, the sol-gel and ultrasonic methods with an immersed sonotrode in a reaction medium were used, followed by heat treatment at low temperature.

The synthesis of nanomaterials with specific electrical and magnetic properties is a field of great interest today. This is due to the particular properties of these materials, being significantly different to those of the bulk material and giving them wide applicability in various fields (including electronics, telecommunications, optics, aerospace, energy, medicine, and biology, etc.). The larger the specific surface area of a nanostructural material, the smaller the particle size. For this reason, the synthesis of nanostructural materials, with particle sizes in the nanometer range (1-100 nm), has attracted the interest of researchers, as evidenced by the multitude of scientific articles published in this field. One of the major current directions of research in this field is the elaboration of new synthesis methods that allow, through accessible technologies, the synthesis of nanomaterials as fine as possible and with controlled morphologies.

In recent years, numerous studies in the field of magnetizable nanosystems, magnetic nanoparticles, and nano-magnetodielectric composites have been undertaken, responding to the current fundamental direction in the development of the nanosciences. The properties of nanocomposites are determined according to their chemical nature, shape, size, the matrix in which the particles are dispersed, and the interaction between the nanoparticles and the matrix.

Metal oxides are one of the best studied categories of materials, as they have a very wide range of structures and properties [1]. The preparation of polycrystalline materials to obtain materials with optimal properties for a desired application can be difficult due to a number of factors that must be taken into account to ensure their reproducibility. When obtaining metal oxide nanoparticles with specific properties by chemical methods, the most important factors to be considered concern their structural properties—crystalline or amorphous structure; size, shape, and morphology of the surface—and chemical properties—composition of the material, interface, and surface.

Metal oxides, having either a spinel, perovskite, or crednerite crystal structure, have been the topic of extensive studies and technological applications due to their useful and varied functional properties (catalytic, optical, electrical, magnetic, etc.), in combination with their chemical, thermal, and mechanical stabilities [2].

The numerous properties displayed by metal oxides, depending on both the structure and nature of the M^1 and M^2 metals, makes this class of materials suitable for applications in an extremely wide variety of fields including medicine, catalysis, oxygen obtaining devices, and electronic and magnetic devices, etc.

In this chapter we present a number of procedures for obtaining metal oxide samples by various methods.

1.2. General Methods for Obtaining Metal Oxides

Numerous methods have been proposed for the synthesis of metal oxides, this diversity being attributed to the difficulties of obtaining particles of specific sizes and with physico-chemical characteristics as close as possible to ideal [3]. Such particle characteristics include:

1. Small-size.
2. Narrow width.
3. Ease of dispersibility.
4. High chemical purity.
5. Homogeneous composition.

The choice of method for preparing metal oxides depends on:

- the raw materials used;
- the preparation scale (laboratory, micro-production, series);
- and the available technical means.

Numerous methods for the synthesis of metal oxides have been suggested. The main methods for obtaining nanocrystalline materials include [4]:

a. Conventional methods (ceramic methods). These consist of obtaining oxide materials in the form of nanoparticles by solid phase reactions at relatively high temperatures from metal oxide precursors, or by thermal decomposition into oxides of carbonates, nitrates, oxalates, and sulphates [5]. The advantage of this method is that we obtain well-crystallized materials. The disadvantage is that the realization of this process involves a series of operations that must be performed before heat treatment that can lead to impurity and loss of material, such as through grinding and mixing [6].
b. Unconventional methods leading to wet oxide nanomaterials, at relatively low temperatures, thus eliminating the specific disadvantages of the ceramic method. The advantage of such wet processes is that we can obtain a homogeneous molecular distribution, ensuring better control of the reaction conditions, with a close dimensional distribution that is almost uniform. In this way, fine-grained particles and a specific and controlled surface can be obtained, depending on the proposed field of application. The degree of crystallization can also be controlled. The most commonly used unconventional methods are: chemical precipitation; the hydrothermal method; the solvothermal method; the sol-gel method; the combustion method; and the microemulsion method [7].

The use of synthetic methods for obtaining metal oxides under conditions of maximum purity and optimum efficiency is justified by the fact that the structure and properties of mixed oxide systems depend mainly on the method used for their synthesis.

In this chapter we present several methods for obtaining metal oxides that have been used by our research team. These methods have been used to obtain manganese ferrite, perovskite, and crednerite-type oxides.

1.3. Synthesis of Manganese Ferrites by Various Processes

This section presents five methods used to obtain manganese ferrite: the classical chemical co-precipitation method; the ultrasonic-assisted co-precipitation method; the direct calcination method; the classical hydrothermal method; and the hydrothermal method with the use of dopants (fig. 1-1).

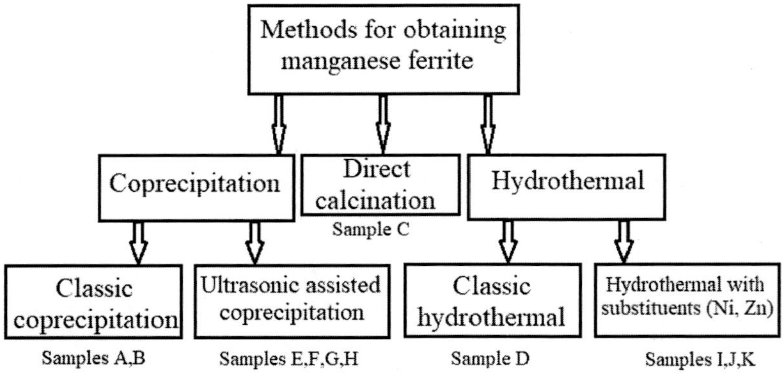

Fig. 1-1. Unconventional synthesis methods for manganese ferrite.

Next, we describe the working technology used for sample preparation, making a comparison with the methods used by other authors for the synthesis of samples of the same type. Images of powdered manganese ferrite samples, obtained by the methods listed above, are shown in fig. 1-2.

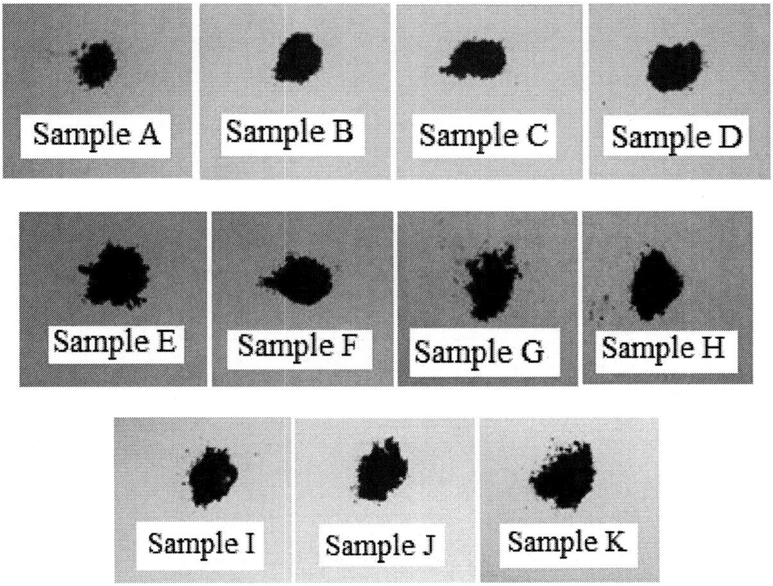

Fig.1-2. Images of manganese ferrite samples obtained.

1.3.1. The classical co-precipitation method

For the synthesis of $MnFe_2O_4$ by the method of co-precipitation, Amighian et al. [8] used an aqueous solution of reactants ($FeSO_4 \cdot 7H_2O$ and $MnSO_4 \cdot H_2O$) and NaOH as a precipitating agent.

In contrast to these authors, who used sulfides as reactants (iron and manganese sulfide), the method proposed here for the preparation by co-precipitation of ferrite powders of $MnFe_2O_4$ (samples A and B) was performed using $MnCl_2{\cdot}4H_2O$ şi $FeCl_3{\cdot}6H_2O$ as reactants. The procedure was as follows: one sample was obtained by simple co-precipitation and calcination at 200 °C and the other by simple co-precipitation and calcination at 900 °C. The samples were synthesized in powder form.

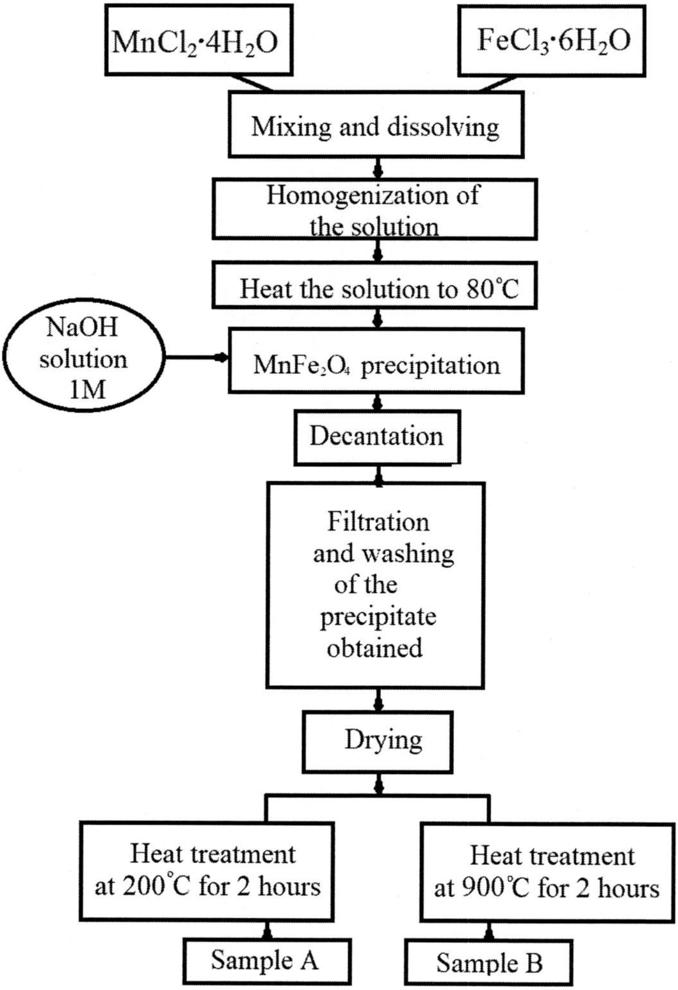

Fig. 1-3. Technology for obtaining manganese ferrite by the method of chemical co-precipitation.

The reagents used to obtain the manganese ferrite powder were mixed in the stoichiometric ratio Mn: Fe = 1: 2 and dissolved in distilled water. The reactants were 0.945 g $MnCl_2 \cdot 4H_2O$, and 2.733 g $FeCl_3 \cdot 6H_2O$ to which 16 ml of 2 M NaOH solution was added until the mixture reached pH = 11. Precipitation of the ferrite was performed by mixing with a Heidolph-type magnetic stirrer equipped with a heating plate for 2 hours at a temperature of 80-90 °C. After decanting, filtering, and washing of the precipitate with ethyl alcohol to remove chlorides and neutralize the solution to pH = 7, the samples were dried in an oven at 80 °C. One sample was then heat treated at 200 °C and the other at a temperature of 900 °C. A diagram of the process for obtaining manganese ferrite by the co-precipitation method is shown in fig. 1-3 [9, 10].

The obtained powder was characterized by X-ray diffraction using the PANalytical X'Pert PRO MRD diffractometer to identify the existing phases and highlight the degree of crystallization. The morphology and structure of the surfaces were studied using a scanning electron microscope (SEM-EDX).

1.3.2. The ultrasonically-assisted co-precipitation method

The principle of this method is the same as for the classical hydrothermal method, but the stirring of the reactant mixture is done with the help of ultrasound, the autoclavation time being 100 minutes. The use of ultrasound as a means of obtaining manganese ferrite nanocrystals has the advantage that crystallization is favoured by the ultrasonic field and takes place at lower temperatures and pressures than in the case of the classical hydrothermal method.

The crystallization rate increases due to the increased number of nucleation centres produced by ultrasonic agitation. The phenomenon of controlled cavitation in the reaction medium minimizes the growth of crystals and the formation of conglomerates so that the obtained particles have a narrow dimensional distribution.

To obtain manganese ferrite, the corresponding metal salts were used as precursors: 2.02 g $Fe(NO_3)_3 \cdot 2H_2O$, 1.0749 g $Mn(NO_3)_2 \cdot H_2O$, and 1 M NaOH solution as the precipitating reagent. All reagents were purchased from Merck and were of confirmed analytical purity. The metal salts were dissolved in distilled water and then the 1 M NaOH solution was added with continuous stirring up to a pH value of 11, with the corresponding metal hydroxides co-precipitating.

To obtain the final product, the suspension obtained was subjected to ultrasound for 20 minutes at 80 °C using a high-intensity ultrasonic processor Sonics VCX-750 Vibra-Cell equipped with a probe that was immersed in the reaction mixture. The working frequency used was 20 kHz, the net power was 750 kW, the amplitude was 80 %, and the on/off pulses were set at 30: 10 (30 seconds running and 10 seconds at rest).

The pulsation cycle was programmed so as not to require additional cooling of the suspension. Due to the replacement of the conventional heating system by an ultrasonic field heating system with an immersed sonotrode, the synthesis time was considerably reduced.

The precipitate obtained was separated by decantation, followed by filtration and washing several times with distilled water to give a neutral pH. Finally, it was washed 2-3 times with ethanol and the washed precipitate was dried in an oven at 80 ° C for 2 hours. From this first sample, three other samples were collected that were subjected to heat treatment at 400 °C, 700 °C, and 1,000 °C [11, 12]. The workflow for the process of obtaining manganese ferrite is presented in fig. 1-4.

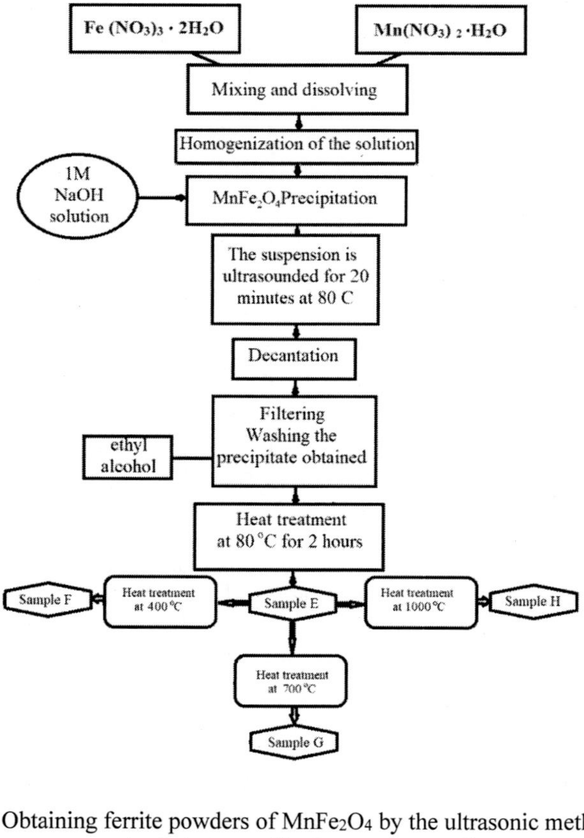

Fig. 1-4. Obtaining ferrite powders of $MnFe_2O_4$ by the ultrasonic method with a sonotrode immersed in the reaction medium.

1.3.3. The direct calcination method

A sample was obtained using $MnCl_2 \cdot 4H_2O$ and $FeCl_3 \cdot 6H_2O$ as precursors, using the same amount of each reactant as in section 1.2.1. Thus, a mixture of 0.945 g $MnCl_2 \cdot 4H_2O$ with 2.733g $FeCl_3 \cdot 6H_2O$ was obtained by grinding and the addition of 0.61 g urea, followed by washing with ethanol. The ethanol was evaporated by heating and the resultant mixture was burnt at 700 °C for 4 hours. The workflow for the process of obtaining manganese ferrite by direct calcination is presented in fig. 1-5 [13].

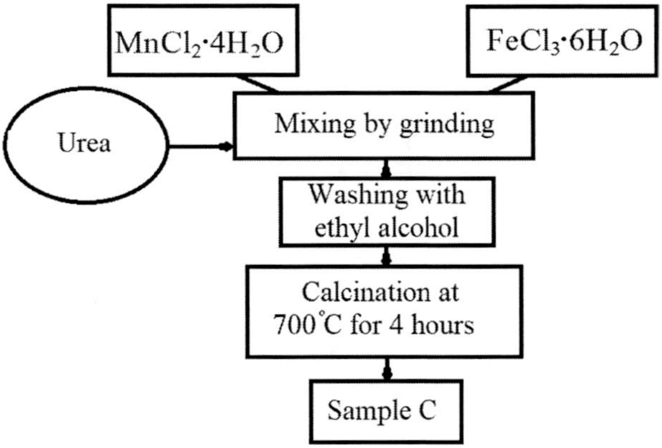

Fig. 1-5. Obtaining manganese ferrite by the direct calcination method.

1.3.4. Synthesis of manganese ferrites by the hydrothermal method

The hydrothermal process is one of the most successful methods for growing crystals of different materials [49]. This method has a number of advantages, including obtaining a pure product with high crystalline homogeneity and control over the shape and size of the particles.

1.3.4.1. The classical hydrothermal method

A sample was obtained by the hydrothermal method. As with the other samples, the same precursors were used with the same amounts: 0.945 g of $MnCl_2 \cdot 4H_2O$ was dissolved together with 2.733 g of $FeCl_3 \cdot 6H_2O$ in distilled water and 12 ml of 2 M NaOH was added to precipitate the manganese ferrite with a pH of 11. The mixture was introduced into a Morey-type autoclave at a temperature of less than 250 °C and a pressure of 80 bar [14, 15]. The resulting product was subjected to filtration, washing, and drying. A diagram of the workflow for obtaining manganese ferrite by the hydrothermal method is presented in fig. 1-6.

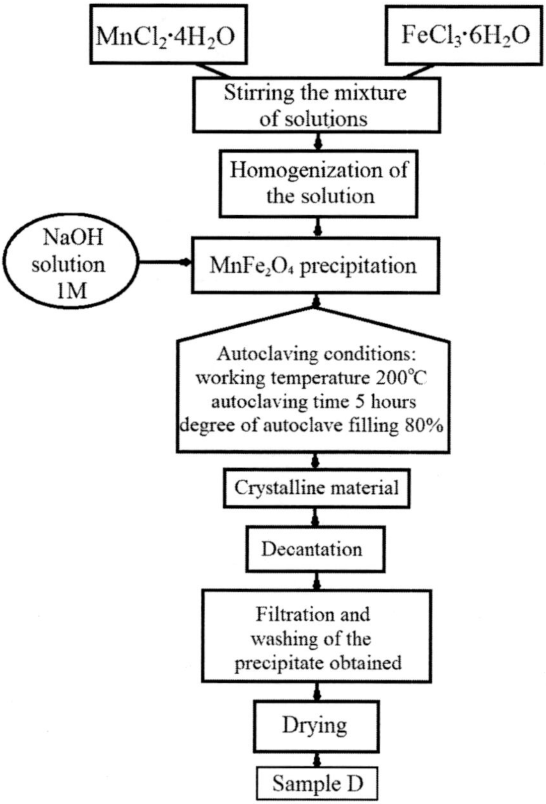

Fig. 1-6. Obtaining manganese ferrite by the simple hydrothermal method

1.3.4.2. The hydrothermal method with Ni and Zn substituents

Hydrothermal manganese ferrite powders and Ni and Zn substituted manganese ferrite powders were obtained. The first samples were obtained by the hydrothermal method using the following reactants: 1.7895 g $Mn(NO_3)_2 \cdot H_2O$, and 4.8372 g Fe $(NO_3)_3 \cdot 2H_2O$. The metal salts were dissolved in distilled water and then 1 M NaOH solution was added to give a pH of 11. The mixture was placed in a Morey-type autoclave at 220 ° C for 12 hours. The resulting product was filtered, washed, and dried to obtain ferrite powders of $MnFe_2O_4$ [16].

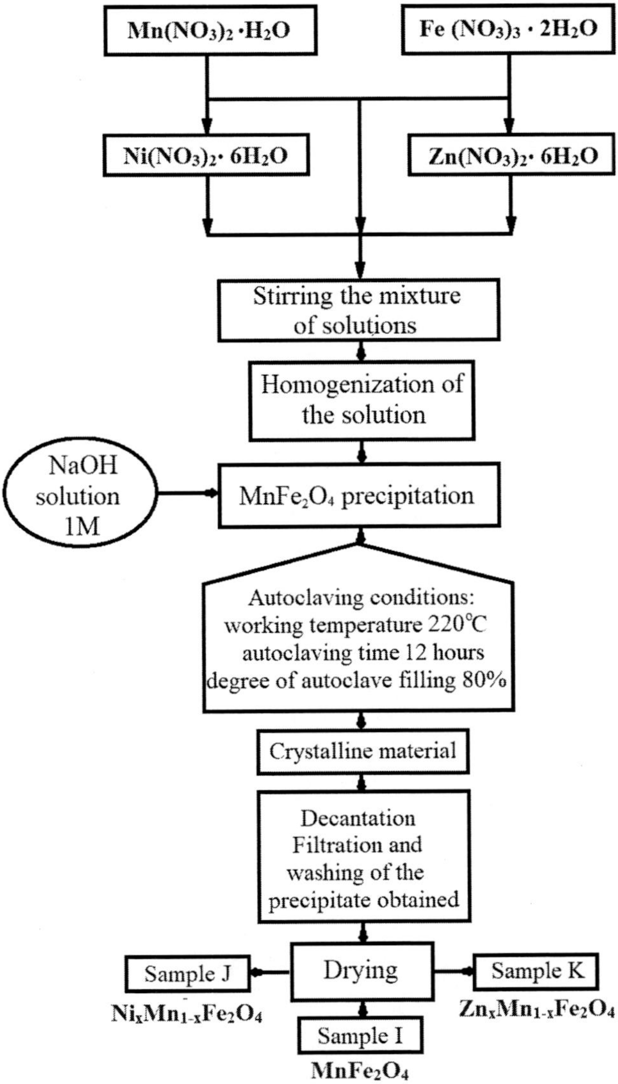

Fig. 1-7. Diagram for obtaining ferrite samples with Ni and Zn substituents.

An Ni-substituted manganese ferrite sample was synthesized from a mixture of three reactants: 1.2554 g $Mn(NO_3)_2{\cdot}H_2O$, 4.8372 g $Fe(NO_3)_3{\cdot}2H_2O$, and 0.5754g $Ni(NO_3)_2{\cdot}6H_2O$ (4 %) solution with a concentration of 0.002 M. The reactants were dissolved in distilled water and then 1 M NaOH solution was added to give a pH value of 14. The mixture was placed in a Morey autoclave at 220 °C for 12 hours. After filtration, washing, and drying, $Ni_xMn_{1-x}Fe_2O_4$ ferrite powder was obtained [17].

A sample of manganese ferrite substituted with Zi was obtained using the following mixture of reactants as precursors: 1.2554 g $Mn(NO_3)_2{\cdot}H_2O$, 4.8372 g $Fe(NO_3)_3{\cdot}2H_2O$, and 0.5949 g $Zn(NO_3)_2{\cdot}6H_2O$ (4 %) solution with a concentration of 0.002 M. The mixture was made by dissolving in distilled water and adding 1 M NaOH solution to give a pH of 12. The mixture was placed in a Morey autoclave at a temperature of 220 °C for 12 hours. After filtration, washing, and drying, $Zn_xMn_{1-x}Fe_2O_4$ ferrite powder was obtained. The workflow for obtaining these ferrite samples is presented in fig. 1-7.

1.4. Obtaining ABO_3-type Compounds with a Perovskite Structure

Here, we present the following proposed methods of synthesis for obtaining $NaTaO_3$ samples: the hydrothermal method; the sol-gel method; and the ultrasound method.

The physical properties of perovskite-type oxides can be greatly modified by changing their crystalline and electronic structures. At the same time, the perovskite structure allows the substitution of metallic elements in the periodic table, either in the tetrahedral state A, or in the octahedral state B [18, 19]. This leads to a very large number of compounds with remarkable properties, including ferroelectric ($BaTiO_3$, $KNbO_3$, PZT), piezoelectric (PZT, (K, Na) NbO_3-NN), and ferromagnetic ($SrRuO_3$), compounds with nonlinear optical ($LiNbO_3$, $LiTaO_3$) and catalytic properties ($NaTaO_3$); compounds with ionic-electronic conductivity ((La, Sr) MnO_3); and compounds with high electronic mobility ($SrTiO_3$) [18, 20].

The synthesis of these materials, with particle sizes in the range of 1-100 nm, has attracted the interest of researchers, especially over the past decade when research began to focus on finding new ceramic materials with thermoelectric properties. Perovskite materials, such as $SrTiO_3$ [21, 22], $NaCoO_3$ [23], and $CaCoO_3$ [24], have been shown to have a high ZT merit factor. Continuing research in this direction, new perovskite materials have been identified and developed, such as sodium tantalum ($NaTaO_3$), with thermoelectric properties, as well as good photocatalytic characteristics [25]. At the same time, the $NaTaO_3$-Fe_2O_3 composite material has a high Seebeck voltage of -300 mV at a high temperature gradient, $\Delta T = 650$ K, and with linear temperature dependence [26, 27, 28]. However, its electrical conductivity is low, so the power factor must be increased.

In this section, we present three methods of synthesis for obtaining $NaTaO_3$ samples: the hydrothermal method; the sol-gel method; and the ultrasonic method. All the $NaTaO_3$ samples obtained were undoped (pure) or doped with Cu or Al metal ions [29].

1.4.1. The hydrothermal method for the synthesis of powdered $NaTaO_3$ samples

The precursors used for the synthesis of $NaTaO_3$ ceramic materials by the hydrothermal method were: 0.500 g of tantalum powder with a purity of 99.9 %, 5 ml of hydrofluoric acid (HF), and 0.5 g of citric acid (as a chelating agent) in 75 ml H_2O and NaOH to adjust the pH to 6.5. The mixture was then introduced into a Morey-type autoclave with a capacity of 37.6 ml, occupying 80 % of the total volume of the autoclave. The autoclave solution was maintained at 160 °C for sample A and 200 °C for 12 hours for sample B.

After decantation and filtration, the resulting precipitate was washed with double-distilled water on filter paper and then dried in an air oven at 800 °C resulting in the two powder samples. In order to obtain new samples of sodium tantalite ($NaTaO_3$) by the hydrothermal method, the working method was modified from the one used to obtain the 2 initial samples. As precursors 0.500 g of tantalum powder with a purity of 99.9 % and 10 ml of fluoric acid (HF) were also used, with citric acid as a chelating agent. Precipitation was performed in an aqueous sodium hydroxide solution (2 M NaOH) at pH = 9 and the solution was stirred with a magnetic stirrer for 2 hours at 85 °C.

The solution obtained was then introduced into the Morey type autoclave and synthesis of the autoclave mixture was done by heat treatment at two temperatures for 6 and 8 hours, respectively, obtaining the following 4 samples: sample S1, 600 °C for 6 h; sample S2, 800 °C for 6 h; sample S3, 600 °C for 12 h; and sample S4, 800 °C for 12 h. For each sample, after decantation and filtration, the resulting precipitate was washed with distilled water on filter paper and then dried in an air oven at 800 °C, giving 4 samples of powdered sodium tantalite [30, 31].

1.4.2. Synthesis of $NaTaO_3$ samples in the form of undoped/doped powders with Al or Cu ions by the ultrasonic method

The precursors used for the synthesis of $NaTaO_3$ ceramic materials, doped or undoped with Cu or Al ions, by the ultrasonic method with a sonotrode immersed in the reaction medium were: tantalum ethoxide ($Ta(OC_2H_5)_5$), sodium hydroxide (NaOH), and copper nitrate ($Cu(NO_3)_2$) or aluminium nitrate ($Al(NO_2)_3$), representing about 0.04 %, all of analytical quality, along with bidistilled water and ethyl alcohol.

The detailed procedure for synthesis is as follows: 1 ml of tantalum ethoxide was added to 50 ml of bidistilled water, stirred continuously by a magnetic stirrer, and with the addition of 0.017 g of copper nitrate or 0.0375 g of aluminum nitrate. Precipitation was performed in an aqueous sodium hydroxide solution (2 M NaOH) up to pH = 13. The obtained suspension was subjected to an ultrasonic process for 15 minutes, with the help of the mobile processor and the sonotrode Sonics Vibra Cell immersed in the reaction medium. Its working parameters were: amplitude 80 %; and pulsation: 10 seconds On/5 seconds Off. The obtained precipitate was filtered and washed with bidistilled water to give a neutral pH and then it was heat treated at a temperature of 600 °C for 6 hours [31]. In this way, by an efficient and simple method using ultrasonic vibrations in the reaction medium, as first reported by us in [32], we obtained the following samples: $NaTaO_3$ doped with aluminium ions (noted US-Al sample) and $NaTaO_3$ doped with copper ions (noted US-Cu sample).

1.4.3. Synthesis of samples of $NaTaO_3$ in the form of undoped/doped powder with Al or Cu ions by the sol-gel method

For the synthesis of sodium tantalite ($NaTaO_3$) samples by the sol-gel method, purely analytical chemicals were used (Aldrich 99.98 %). The initial materials used were 1 ml of tantalum ethoxide and a solution (1: 1) of ethyl alcohol and distilled water at room temperature. Precipitation was performed in an aqueous sodium hydroxide solution (1 M NaOH) at pH = 7, with continuous stirring using a magnetic stirrer. The resultant mixture was a white suspension that was heated for a further 2 hours at 90 °C.

A gel was obtained that had been dried in an oven at 110 °C for 4 hours. For crystallization, the precipitate was heat treated for 6 hours at 600 °C, with a heating rate of 5 °C. A sample was obtained by the sol-gel method of pure $NaTaO_3$ in powder form, denoted the SG sample. For the synthesis of sodium tantalate ($NaTaO_3$) doped with Al or Cu metal ions, the same precursors and the same reaction conditions were used, but 0.04 g of copper nitrate with the formula $Cu(NO_3)_2$ were added to the suspension with continuous stirring, or aluminum nitrate with the formula $Al(NO_3)_3$. The resulting suspension was heated for 2 hours at 90 °C and the gel-like product was then dried in an oven at 110 °C for 4 hours. Finally, for crystallization, the $NaTaO_3$ samples, doped with Al ions or Cu ions, were subjected to heat treatment for 6 hours at a temperature of 600 °C, with a heating rate of 5 °C/min [33, 34].

1.5. Synthesis of Crednerite by Various Processes

The specialized literature presents varied and complex synthesis methods for obtaining $CuMnO_2$-type materials as single-phase polycrystalline materials, thin films, and single crystals.

For the production of polycrystalline materials of the crednerite type, synthesis was by the sol-gel method [9], with preparation at low temperatures by exchange reactions [35], or by solid state reactions at high temperatures (950 °C) starting from the precursors CuO and Mn_2O_3 in closed quartz tubes under vacuum [36] with an atmosphere of Ar [37] or N_2 [38]. The sol-gel method was used for the preparation of thin films [39] and $Cu_{1.04}Mn_{0.96}O_2$ crystals were obtained by the flux metho [40]. In the synthesis of the $CuMnO_2$ compound at low temperature and pressure by the hydrothermal method [41-43], in the case of obtaining delafossite-type materials based on copper, one of the main difficulties is the stabilization of Cu^+ species in hydrothermal conditions. In aqueous solution at room temperature, Cu^+ is not stable at any pH value and dissociates into Cu^o and Cu^{2+}.

$Cu_1 + xMn_{1-x}O_2$ crednerite-type compounds (with x = 0 and 0.06) were prepared by the solid state reaction method, as described in [44, 45, 46]. Oxides of CuO and MnO were mixed and ground together, with the mixture obtained having the stoichiometric ratio (1 + x) CuO: (1-x) MnO, as shown in fig. 1-8. The compounds obtained (for x = 0 and x = 0.06, respectively) were pressed to give pellet-shaped samples (1 cm in diameter and a few millimetres thick). The pellets were then placed in vacuum-evacuated silicon tubes and heated in an oven at 950 ° C for 12 hours [47].

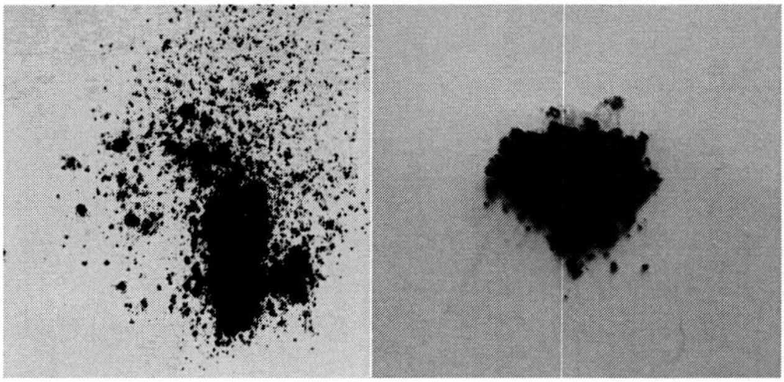

Fig. 1-8. $Cu_{1+x}Mn_{1-x}O_2$ powder-type crednerite samples

1.5.1. Synthesis of materials under vacuum in sealed silica tubes

The synthesis of materials can be performed in a controlled atmosphere of Ar, Ar/H_2, N_2, CO/CO_2, or under vacuum conditions in order to control the stoichiometry and size/shape or obtain single phase compounds. For the synthesis of materials in a vacuum, several approaches can be taken, such as synthesis in a vacuum furnace, or by using sealed silica tubes, as described in the following.

In the first step, the powdered precursors are weighed in stoichiometric proportions, mixed by grinding in an agate mortar, and pressed into pellets (1 cm in diameter and a few millimetres thick) or parallelepiped bars (dimensions: 2 x 2 x 10 mm^3). The compression of the powder induces a good diffusion of species during thermal treatment and this step is required otherwise the vacuum inside the tube would not be possible due to the fact that the powder would be drawn into the vacuum pump. The bars or pellets are then placed in an alumina finger, preventing their diffusion in the silica of the powder during synthesis. The alumina finger is then placed in a quartz tube, which is sealed with a vacuum of 10^{-3} mbar. The sealed silica tubes are then placed in a furnace for thermal treatment, performed in a specific sequence for each synthesis (fig. 1-9).

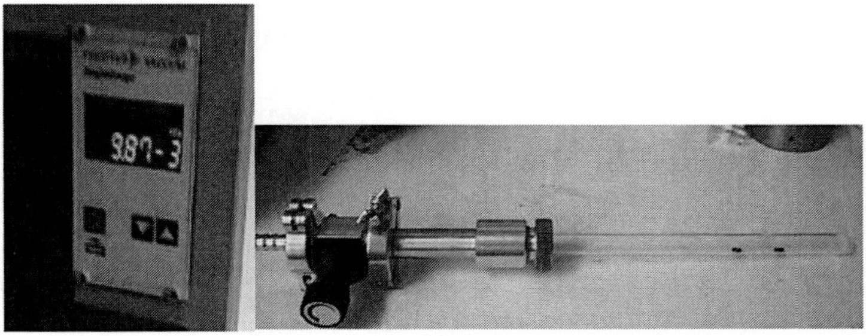

(a)

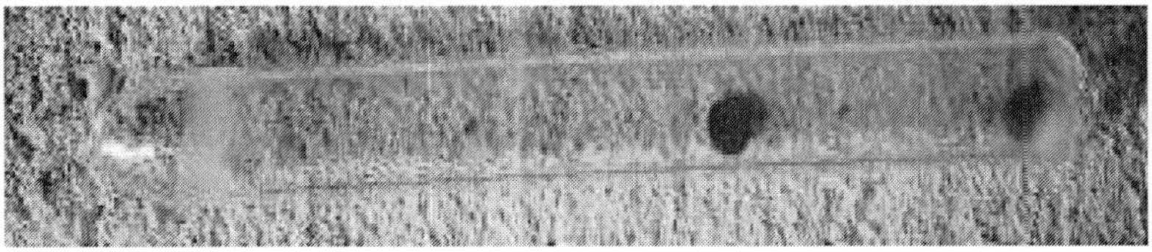

(b)

Fig. 1-9. (a) The system used to evacuate the silica tubes under vacuum and (b) a sealed silica tube with 2 pellets inside (without the alumina finger).

Today, researchers are using this synthesis method for a large variety of materials. Controlled oxygen stoichiometry in the compounds is essential making this a very useful synthesis method for crednerites [48, 49], tungsten bronzes [50], and perovskites [51], for example. For some materials that support a range of chemical compositions, special attention has to be given to the oxygen stoichiometry and sealed silica tubes are used [48].

Polycrystalline $Cu_{1+x}Mn_{1-x}O_{2+\delta}$ (x = 0-0.10)-type materials have been reported in the literature, prepared by solid state reactions in closed quartz tubes under vacuum [48, 49]. In a typical synthesis, CuO and MnO powders were weighed in the stoichiometric proportions 1 + x CuO and 1-x MnO, ground, and pressed into pellets (1 cm in diameter and a few millimetres thick). These were then placed in quartz tubes sealed under primary vacuum and heated in an oven at 950 °C for 12 hours [48]. Solid state reactions in sealed silica tubes have also been carried out starting with stochiometric amounts of Cu, CuO, and Mn_2O_3 in [49] at temperatures ranging from 960 °C for $CuMnO_2$ to 1,020 °C for $Cu_{1.2}Mn_{0.8}O_2$.

1.5.2. The ultrasound-assisted co-precipitation method

In recent years, sonochemistry or ultrasonic cavitation chemistry has emerged as a potential alternative for the synthesis of materials like, for example, ferrites of nano-dimensions [52-56], hydroxyapatite [57], CuO [58], spinels [59], and vaterite-type calcium carbonate [60], etc. This synthesis method is environmentally friendly and allows us to obtain nanosized materials with uniform size distributions that are single phased [61]. In the case of $Ni(OH)_2$, it has been shown to influence the crystal structure [62].

Sonication of a liquid result in rapid heating, driving cavitation, which involves the implosive collapse of bubbles [63]. The relevant chemical reactions largely take place inside these bubbles [64, 65], which otherwise would require rigorous stirring [66]. In hot spots, where the bubbles are heated intensely, the temperatures are roughly 5,000 °C with pressures of about 500 atmospheres and lifetimes of a few microseconds [64, 67-68].

Fig. 1-10. Ultrasonic device consisting of: 1) sonotrode with a titanium tip; 2) ultrasonic processor; and 3) temperature controller.

In the process of synthesis, the time, temperature, and amplitude of the ultrasound is tested in order to find the best parameters to obtain the desired materials. A typical experimental strategy consists of dissolving the precursors (acetates, nitrates, etc.) in distilled water before carrying out sonication for several minutes (10, 20, 30… min). Sonication is applied using an ultrasonic probe (1/2in), as shown in fig. 1-10 (Sonics Vibra Cell).

$CuMnO_2$ nanocrystals were prepared by the ultrasound-assisted co-precipitation method for the first time in [69]. The method consisted of dissolving stoichiometric quantities of Cu $(NO_3)_2{\cdot}3H_2O$ and Mn $(NO_3)_2{\cdot}4H_2O$ in a 2.5 M NaOH solution. The mixing of both solutions determined the precipitation, after being subjected to ultrasonic vibration for set periods of time (20, 30, or 40 min) at 80 °C, a frequency of 20 kHz, a pulsation amplitude of 80 %, and a pulsation cycle of 10 s ON/5 s OFF.

1.6. Conclusion

The study of metal oxides is currently a major research direction, both in terms of using new methods of synthesis and the comparative study of the electrical and magnetic performance of samples obtained for use in various applications.

The research conducted and presented in this chapter involved the use of different methods for obtaining metal oxides, including: the classical chemical co-precipitation method; the ultrasonic-assisted co-precipitation method; the direct calcination method; and variants of the hydrothermal method (classical hydrothermal and hydrothermal with the use of dopants). Through the numerous methods used, single-phase or multiphase powder samples were obtained.

These oxide materials samples were analysed both structurally and in terms of surface morphology by X-ray diffraction (XRD), scanning electron microscopy/energy-dispersive X-ray spectroscopy (SEM/EDX), UV-Vis spectroscopy, Fourier transform infrared spectroscopy (FTIR), and thermogravimetric analysis (TGA).

The investigation of the electrical and magnetic properties of the studied nanomaterials was performed to give electrical and magnetic measurements using complex impedance spectroscopy, complex dielectric permittivity, and thermoelectric analysis of the complex conductivity, along with the inductive method of graphing the hysteresis.

References

[1] Iu. E. Levin, Iu. D.Tretiakov, L. M. Letiuc, Fizico-himiceskie osnovi poluceniia
svoistva i primenenie ferritov, Metallurghia, Moskva (1979).
[2] G. Winkler, Crystallography, chemistry,and technology of ferrites, in: Magnetic Properties of Materials, ed. J. Smit, McGraw-Hill Book Comp., N. Y. (1971) 20.
[3] R. Valenzuela, Magnetic Ceramics, Cambridge Univ. Press (1994).
[4] A. Goldman, Modern Ferrite Technology, Springer N. Y. (2006).
[5] Zhang De Yuan, Zang Wen Qiang, Magnetization of microorganism cells by thermal decomposition method, *Tehnological Sciences* 5 (2011) 1275.
[6] V. T. Zaspalis, E. S. Kikkinides, M. Kolenbrander, R. Mauczok, Method for the morphological characterization of powder raw materials for the manufacturing of ceramics, *J. Mat. Proc. Tech.* vol. 142 (2003) 267–274.
[7] T. A. Ring, Fundamentals of Ceramic Powder Processing and Synthesis, *Acad. Press*, San Diego (1995).
[8] Amighian, J., Mozaffer, M., and Nasr, B. Preparation of nano-sized manganese ferrite (MnFe2O4 via coprecipitation method.) *Physica Status Solidi. C, Current Topics in Solid State Physics* 3 (2006) 3188–3192.
[9] A. Lungu, I. Malaescu, C. N. Marin, P. Vlazan, P. Sfirloaga, The electrical properties of manganese ferrite powders prepared by two different methods, *Physica B* 462 (2015) 80–85.
[10] Y. M. Z. Ahmed, Synthesis of manganese ferrite from non-standard raw materials using ceramic technique, *Ceramics International* 36 (2010) 969–977.
[11] I. Mălăescu, A. Lungu, C. N. Marin, Paulina Vlăzan, Paula Sfirloagă, G. M. Turi, Experimental investigations of the structural transformations induced by heat treatment in manganese ferrite synthesized by ultrasonic-assisted co-precipitation method, *Ceramics International* 42 (2016)16744–16748.
[12] A. Lungu, P. Vlazan, I. Malaescu, C. N. Marin, P. Sfirloaga, Electric properties of manganese ferrite obtained by different methods, TIM 13 Physics Conference, November 21-23, (2013) Timisoara, Romania.
[13] I. Mălăescu, A. Lungu, C. N. Marin, Paulina Vlăzan, Paula Sfirloagă, Complex impedance of manganese ferrite powders obtained by two different methods, *Analele Universității de Vest din Timişoara*, Seria Fizică, vol. LVIII, (2015) 14–26 (DOI: 10.1515/awutp -2015-0203).
[14] Yin M. and O'Brien S., Synthesis of Monodispersed Nanocrystals of Manganese Oxides *J. Am. Chem. Soc.* 125 (2003) 10180-10181.
[15] Ravindranathan P and Patil K. C., A low temperature path to ultrafine ferrites *Bull. Amer. Ceram. Soc.* 66 (1987) 688–692.
[16] D. Makovec, M. Drofenik, A. Žnidaršic, Sintering of MnZn-ferrite powders prepared by hydrothermal reactions between oxides, *J. Eu .Ceram. Soc.* 21 (2001) 1945–1949.
[17] S. Komarneni, E. Fregeau, E. Breval, R. Roy, Hydrothermal preparation of ultrafine ferrites and their sintering, *J. Am. Ceram. Soc.* vol. 71 (1988) C-26–C-28.

[18] D. G. Schlom, L.-Q. Chen, X. Pan, A. Schmehl, M. A. Zurbuchen, A Thin Film Approach to Engineering Functionality into Oxides, *J. Am. Ceram. Soc.* 91 (2008) 2429.

[19] F. Wang, I. Grinberg, A. M. Rappe, Band Gap Engineering Strategy via Polarization Rotation in Perovskite Ferroelectrics, *Appl. Phys. Lett.* 104 (2014) 152903.

[20] V. Fedorov, Perovskites, in: Ceramics Science and Technology, Wiley-VCH Verlag GmbH & Co. KGaA, Weinheim (2010) 257–297.

[21] S. Ohta, T. Nomura, H. Ohta, H. Hosono, K. Koumoto, Large thermoelectric performance of heavily Nb-doped SrTiO3 epitaxial film at high temperatures, *Appl. Phys. Lett.* 87 092108 (2005).

[22] H. Ohta, S. Kim, Y. Mune, T. Mizoguchi, K. Nomura, S. Ohta, T. Nomura, Y. Nakanishi, Y. Ikuhara, M. Hirano, H. Hosono, K. Koumoto, Giant thermoelectric Seebeck.

[23] I. Terasaki, Y. Sasago, K. Uchinokura, Large thermoelectric power in NaCo2O4 single crystals, *Phys. Rev. B* 56 R12685 (1997).

[24] R. Funahashi, I. Matsubara, H. Ikuta, T. Takeuchi, U. Mizutani, S. Sodeoka, A single oxide crystal with high thermoelectric performance in air, *Jpn. J. Appl. Phys.* 39 L1127 (2000).

[25] A. Kudo, and H. Kato, Effect of lanthanide-doping into NaTaO3 photocatalysts for efficient water splitting, *Chemical Physics Letters* 331 (2000) 373–377.

[26] W. Wunderlich, NaTaO3 composite ceramics: A new thermoelectric material for energy generation, *J. Nuclear Mater.* 389 (2009) 57–61.

[27] W. Wunderlich, T. Mori, O. Sologub, B. Baufeld, SPS-Sintering of NaTaO3-Fe2O3 Composites, *J. Australian Ceramic Society* 47, 2 (2011) 57–60.

[28] Wilfried Wunderlich, Susumu Soga, Microstructure and Seebeck voltage of NaTaO3 composite ceramics with additions of Mn, Cr, Fe or Ti, *Journal of Ceramic Processing Research*, vol. 11, No. 2 (2010) 233–36.

[30] D. Malaescu, I. Grozescu, P. Sfirloaga, P. Vlazan, C. N. Marin, Electric properties of NaTaO3 obtained by the hydrothermal method, Analele Universităţii de Vest din Timişoara, Seria Fizică, vol. LVIII (2015) 48–55.

[31] P. Sfirloaga, I. Malaescu, M. Poienar, M. C. Nicolae, D. Malaescu, P. Vlazan, Synthesis, structural and electrical properties of NaTaO3:Cu, *J. Mater. Sci: Mater. Electron.* 27, 11 (2016) 11640–11645, DOI 10.1007/s10854-016-5299-2.

[32] D. Malaescu, I. Grozescu, P. Sfirloaga, P. Vlazan, C. N. Marin, The Electrical Properties of Some Composite Materials Based on Sodium and Tantalum Oxides, *Acta Physica Polonica A* vol. 129, Nr. 1 (2016) 133–137.

[33] P. Sfirloaga, I. Malaescu, M. Poienar, M. C. Nicolae, D. Malaescu, P. Vlazan, Synthesis, structural and electrical properties of NaTaO3:Cu, *J. Mater. Sci: Mater. Electron.* 27, 11 (2016) 11640–11645, DOI 10.1007/s10854-016-5299-2.

[34] D. Malaescu, I. Grozescu, P. Sfirloaga, P. Vlazan, C. N. Marin, The Electrical Properties of Some Composite Materials Based on Sodium and Tantalum Oxides, *Acta Physica Polonica A* vol. 129, Nr. 1 (2016) 133–137.

[35] Yassine Bessekhouad, Yamina Gabes, Aissa Bouguelia, Mohamed Trari, The physical and photo electrochemical characterization of the crednerite CuMnO2, *J Mater. Sci.* 42 (2007) 6469–6476.

[36] F. Damay, M. Poienar, C. Martin, A. Maignan, J. Rodriguez-Carvajal, G. André and J. P. Doumerc, *PHYSICAL REVIEW B* 80 (2009) 094410.

[37] B. Bellal, B. Hadjarab, N. Benreguia, Y. Bessekhouad, M. Trari, *J. Appl. Electrochem.* (2011) 41, 867–872.

[38]. S. Kato, R. Fujimaki, M. Ogasawara, T. Wakabayashi, Y. Nakahara and S. Nakata, *Appl. Catalysis B: Environmental* 89 (2009) 183.

[39] Hong-Ying Chen, Da-Je Hsu, *Applied Surface Science* 290 (2014) 161–166.

[40] J. Töpfer, M. Trari, P. Gravereau, J. P. Chaminade and J. P. Doumerc, *Zeit. Kristallogr.* 210, 184 (1995).

[41] William C. Sheets, Emmanuelle Mugnier, Antoine Barnabe, Tobin J. Marks, and Kenneth R. Poeppelmeier, *Chem. Mater.* 2006, 18, 7–20.

[42] M. Poienar, R. Banica, P. Sfirloaga, C. Ianasi, C. V. Mihali, P. Vlazan, *Ceramics International* 44 (2018) 6157-6161.

[43] M. Poienar, P. Sfirloaga, C. Martin, D. Ursu, P. Vlazan, *Journal of Materials Science* 53 (2018) 2389–2395.

[44]D. Kondrashev, The crystal structure and composition of crednerite, CuMnO2, *Soviet Physics: Crystallography*, 3 (1959) 703-706.

[45] F. Damay, M. Poienar, C. Martin, A. Maignan, J. Rodriguez-Carvajal, G. André, J. P. Doumerc, Spin-lattice coupling induced phase transition in the S=2 frustrated antiferromagnet CuMnO2, *Phys. Rev. B*. 80 (2009) 094410-7.

[46] M. Poienar, C. Vecchini, G. André, A. Daoud-Aladine, I. Margiolaki, A. Maignan, A. Lappas, L. Chapon, M. Hervieu, F. Damay, C. Martin, Substitution Effect on the Interplane Coupling in Crednerite: the Cu1.04Mn0.96O2 Case, *Chem. Mater.* 23(1) (2011) 85–94.

[47] I. Malaescu, A. Lungu, C. N. Marin, P. Sfirloaga, P. Vlazan, S. Brindusoiu, M. Poienar, Temperature dependence of the dynamic electrical properties of Cu1+xMn1-xO2 (x=0 and 0.06) crednerite materials, *Ceramics International* 44(10) (2018) 11610–11616.

[48] I. Malaescu, A. Lungu, C.N. Marin, P. Sfirloaga, P. Vlazan, S. Brindusoiu, M. Poienar, Temperature dependence of the dynamic electrical properties of Cu $_{1+x}$ Mn $_{1-x}$ O 2 (x=0 and 0.06) crednerite materials, *Ceramics International*, volume 44, Issue 10, July 2018, pp. 11610-11616.

[49] M. Trari, J. Töpfer, P. Dordor, J. C. Grenier, M. Pouchard, J. P. Doumerc, Preparation and physical properties of the solid solutions Cu1+xMn1−xO2, *Journal of Solid State Chemistry*, volume 178, Issue 9, September 2005, Pages 2751–758.

[50] $K_xV_yW_{1-y}O_3$, A. M. Abdullah, T. Debnath, C. H. Rüscher, and A. Hussain, Synthesis and Characterization of Vanadium Substituted Potassium Tungsten Bronzes, *J. Sci. Res.* 4 (2), 507–514 (2012).

[51] Tapas Kumar Mandal, Mark Croft, Joke Hadermann, Gustaaf Van Tendeloo, Peter W. Stephens and Martha Greenblatt, La_2MnVO_6 double perovskite: a structural, magnetic and X-ray absorption investigation, *J. Mater. Chem.* 2009, 19, 4382–4390.

[52] M. Sivakumar, A. Gedanken, W. Zhang, Y.W. Du, D. Bhattacharya, Y. Yeshurun, I. Felner, Nanophase formation of strontium hexaferrite fine powder by the sonochemical method using $Fe(CO)_5$, *J. Magn. Magn. Mater.* 268 (2004) 95.
[53] M. Sivakumar, A. Gedanken, D. Bhattacharya, I. Brukental, Y. Yeshurun, W. Zhang, Y.W. Du, I. Felner, I. Nowik, Sonochemical Synthesis of Nanocrystalline Rare Earth Orthoferrites Using $Fe(CO)_5$ Precursor, *Chem. Mater.* 16 (2004) 3623.
[54] M. Sivakumar, A. Gedanken, W. Zhang, Y. H. Jiang, Y. W. Du, I. Brukental, D. Bhattacharya, Y. Yeshurun, I. Nowik, Sonochemical synthesis of nanocrystalline $LaFeO_3$, *J. Mater. Chem.* 14 (2004) 764.
[55] M. Sivakumar, T. Takami, H. Ikuta, A. Towata, K. Yasui, T. Tuziuti, T. Kozaka, D. Bhattacharya, Y. Iida, Fabrication of Zinc Ferrite Nanocrystals by Sonochemical Emulsification and Evaporation: Observation of Magnetization and Its Relaxation at Low Temperature, *J. Phys. Chem. B* 110 (2006) 15234.
[56] M. Sivakumar, A. Towata, K. Yasui, T. Tuziuti, Y. Iida, *Curr. Appl. Phys.* 6 (2006).
[57] Dhiraj Mehta, Poonam Mondal, Virendra Kumar Saharan, Suja George, Synthesis of hydroxyapatite nanorods for application in water defluoridation and optimization of process variables: Advantage of ultrasonication with precipitation method over conventional method, *Ultrasonics Sonochemistry* 37 (2017) 56–70.
[58] Purushottam Kumar Singh, Alok Kumar Das, Goutam Hatui, Ganesh Chandra Nayak, Shape controlled green synthesis of CuO nanoparticles through ultrasonic assisted electrochemical discharge process and its application for supercapacitor, *Materials Chemistry and Physics* 198 (2017) 16–34.
[59] Ting-Feng Yi ∗, Xin-Guo Hu, Preparation and characterization of sub-micro $LiNi_{0.5-x}Mn_{1.5+x}O_4$ for 5V cathode materials synthesized by an ultrasonic-assisted co-precipitation method, *Journal of Power Sources* 167 (2007) 185–191.
[60] J. A. Juhasz-Bortuzzo, B. Myszka, R. Silva, A. R. Boccaccini, Sonosynthesis of Vaterite-Type Calcium Carbonate, *Cryst. Growth Des.* 17(5) (2017), 2351–2356. https://doi: 10.1021/acs.cgd.6b01493.
[61] J. H. Bang, K. S. Suslick. Applications of ultrasound to the synthesis of nanostructured materials, *Adv. Mater.* 22 (2010) 1039–1059. https://doi.org/10.1002/adma.200904093.
[62] S. Cabanas-Polo, Z. Gonzalez, A. J. Sanchez-Herencia, B. Ferrari, Influence of ultrasound in the instantaneous synthesis of tridimensional α-$Ni(OH)_2$ nanostructures and derived NiO nanoparticles, *Cryst. Eng. Comm.* 17 (2015) 6193–6206, doi: 10.1039/c5ce00876j.
[63] K. S. Suslick, Sonochemistry, *Science* vol. 247 (1990) 1439–1445, doi: 10.1126/science.247.4949.1439.
[64] Brian L. Cushing, Vladimir L. Kolesnichenko, and Charles J. O'Connor, Recent Advances in the Liquid-Phase Syntheses of Inorganic Nanoparticles, *Chem. Rev.* 2004, 104, 3893–3946.

[65] V. S. Moholkar, S. P. Sable, A. B. Pandit, Mapping the cavitation intensity in an ultrasonic bath using the acoustic emission, *AIChE J.* 46 (2000) 684–694.
[66] E. B. Flint, K. S. Suslick. The Temperature of Cavitation, *Science* 253 (1991) 1397.
[67] A. A. Atchley, L. A. Crum. In *Ultrasound*, Suslick, K. S. (ed.), VCH: New York (1988).
[68] K. S. Suslick, D. A. Hammerton, R. E. J. Cline. Sonochemical hot spot, *Am. Chem. Soc.* (1986) 108, 5641.
[69] Maria Poienar, Antoanetta Lungu, Paula Sfirloaga, Mihai Lungu, Ciprian Valentin Mihali and Paulina Vlazan. Use of ultrasound-assisted co-precipitation route to obtain CuMnO2 semiconductor nanomaterials, *Chemical Papers* vol. 73 (2019) 1541–1546.

CHAPTER 2

METHODS OF MORPHO-STRUCTURAL CHARACTERIZATION AND DETERMINATION OF THE ELECTRICAL AND MAGNETIC PROPERTIES OF METAL OXIDES

IOSIF MALAESCU, AUREL ERCUȚA, PAULA SFIRLOAGĂ, MARIA POIENAR, PAULINA VLAZAN, ANTOANETTA LUNGU, CATALIN N. MARIN, MIHAIL LUNGU

This chapter summarizes methods of structural and morphological analysis that can be applied to metal oxide materials and techniques for determining their electrical and magnetic properties. From among the large range of characterization methods available (morphological, structural, elementary, electrical, dielectric, and magnetic etc.), we present those methods that have been used in the characterization of synthetized metal oxides in our research. Structural and morphological analysis of materials was performed primarily by X-ray diffraction (XRD), scanning electron microscopy/spectroscopy energy-dispersive X-ray (SEM/EDX), UV-Vis spectroscopy, Fourier transform infrared spectroscopy (FTIR), and thermogravimetric analysis (TGA). We present the results obtained from our analyses of different oxide materials. The investigation of the electrical and magnetic properties of the studied nanomaterials was performed using various electrical and magnetic measurements, including: complex impedance spectroscopy; complex dielectric permittivity; thermoelectric analysis of complex conductivity; and the inductive method of the hysteresis graph.

2.1. Introduction

Metal oxides constitute an important class of materials with a wide variety of chemical compositions, crystalline forms, and atomic structures. In general, metal oxides are in the form of a (solid) powder with metallic cations and oxygen anions. These oxides have many physical, optical, electrical, and chemical properties, with uses in making energy storage devices, catalysts, gas sensors, solid state microbatteries, electronic components, optoelectronic devices, and transducers. Given their physical properties, metal oxides can be classified into nontransition metal oxides and transition metal oxides [1, 2]. For example, in the category of nontransition metal oxides we find magnesium oxide (MgO) and silicon dioxide (SiO_2). Under normal conditions, they behave like diamagnetic insulators, but at high temperatures, ionic conduction predominates [3]. In the category of transition metal oxides we have vanadium pentoxide (V_2O_5), tantalum oxide (Ta_2O_5), zirconium dioxide (ZrO_2), and niobium oxide (Nb_2O_5), in all of which the metal ion has a filled *d* electronic configuration [4, 5]. Other metal oxides in this category, such as molybdenum dioxide (MoO_2), tungsten dioxide (WO_2), and titanium monoxide (TiO), are characterized by the fact that the metal ion has a *dx* partially-filled electronic configuration. In recent years, metal oxides and powders, as well as their composites, have been intensively studied due to their novel characteristics and applications [6, 7, 8].

Obtaining and characterizing the morphological, structural electrical, and magnetic properties of metallic oxide materials is necessary for technological processes, research, and many applications [9]. In recent decades, numerous techniques have been developed for the investigation of their crystalline structure, morphology, surface analysis, chemical composition, and optical, electrical, and magnetic properties [10, 11]. Today, instrumental analytical processes are being used more and more, compared to the classic chemical methods, because such analytical methods, have the advantage of being fast, fully automated, and can utilise modern computing power in order to process the experimental data.

Due to the many methods of characterization of metallic oxide materials (morphological, structural, elementary, phase, electrical, dielectric, and magnetic), in this chapter we will only present those methods that have been used in our research [12-15]. In our research, the structural and morphological analysis of materials has mainly been realized by electron microscopy, including SEM, nanoparticle tracking analysis (NTA, to track Brownian motion), powder X-ray diffraction (XRD), Fourier transform infrared spectroscopy (FTIR), spectroscopie UV-VIS, and thermogravimetric analysis (TG/DTG). The investigation of the electrical and magnetic properties of metallic oxide materials was performed using complex impedance spectroscopy measurements, thermoelectric analysis of complex dielectic permittivity, and the inductive method of the hysteresis graph. All the electrical and magnetic measurements were performed on experimental setups in the laboratory, which will be described in this chapter, along with some of the results obtained by us for various metallic oxide samples [16-19].

This chapter has two parts: a) the first part summarizes the methods of structural and morphological analysis that can be applied to metallic oxide samples, along with some examples of analysis; b) the second part describes the laboratory setups and techniques for determining the electrical, dielectric, and magnetic properties of these oxide materials. The experimental studies undertaken into the electrical and magnetic properties of the investigated samples and the analysis of conduction mechanisms in them, as well as their explanation, relied on a number of different theoretical models from the literature.

2.2. Structural Analysis by X-ray Diffraction

In 1912, Max von Laue observed that a crystal can diffract X-rays, thus confirming the hypothesis that this form of radiation has an undulating nature. In 1913, H. W. Bragg and his son W. L. Bragg interpreted X-ray diffraction in relation to crystal lattice planes. Max von Laue assumed that if the atoms in a crystal form spatial networks, the crystal must behave like a diffraction lattice and display characteristics similar to when light is passed through an optical diffraction grating. For this phenomenon to take place, the radiation must have a wavelength comparable to the distance between two neighboring atoms in the network. X-ray diffraction analysis is a non-destructive technique that provides information on the chemical composition and crystalline structure of materials in powder or solid form. Qualitative phase analyses using X-ray spectra are particularly important for the identification of crystalline compounds. Through the diffraction of X-rays by crystalline powders, characteristic spectra specific to each crystalline species can be obtained. These spectra can be used to determine the presence or absence of a crystalline phase [20, 21].

The K_α line is used for the study of X-ray reactive materials. It is commonly filtered with an absorbent film (for example, thin nickel foil). Copper is a widely used metal that can be easily stored at low temperatures because it has a high thermal conductivity and produces strong K_α and K_β lines [21]. The wavelength corresponding to the K_α line of copper is $\lambda = 0.1541$ nm. The basic principle of the method of characterizing materials by X-ray diffraction is based on the connection between the scattering of X-rays and the ordered arrangement of atoms in space. We assume a system of lattice planes, N1, N2, N3, etc., repeated at the same distance *d*. A beam of X-rays falls onto these planes under the angle θ and with wavelength λ (fig. 2-1). The distance *d* between the centers of two neighboring atoms, more precisely its order of magnitude, represents the atomic diameter itself. As a result, this method allows us to measure the size of *d* and thus make it possible to determine the arrangement of the atoms in a material. The interplanar distance *d* is computed using the expression:

$$d = \frac{a_0}{\sqrt{h^2 + k^2 + l^2}} \tag{2.1}$$

where a_0 is the parameter of the crystalline lattice and *h*, *k*, *l* are the Miller indices for different crystalline planes. In a perfect crystal all molecules have the same conformation and orientation.

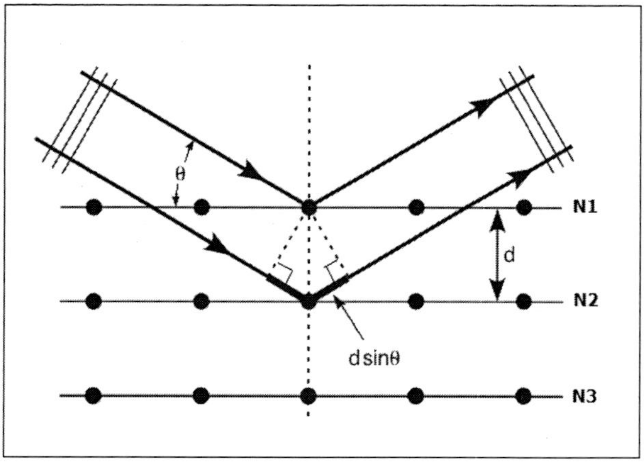

Fig. 2-1. X-ray diffraction of a beam by a series of lattice planes with distance *d* between them.

When X-rays are focused on the crystal, the rays are deflected by the atoms of the molecules in certain directions, causing a particular diffraction pattern. The number of electrons in the atom determines the intensity of the scattered rays. Diffraction spectra obtained from different angles provide information about the components and distribution of atoms in the molecule. Practically speaking, the spectrum resulting from X-ray diffraction analysis (XRD) is obtained as a "reflection" of an X-ray beam from a family of parallel and equidistant atomic particles, due to Bragg's law and given by the relation [22]:

$$n\lambda = 2d\sin\vartheta \tag{2.2}$$

where λ is the wavelength of the X-ray radiation, n is the order of diffraction, and θ is the diffraction angle. Bragg's law is a consequence of the periodicity of the network and is derived from a much more general equation—the Laue diffraction equation [23].

The phase analysis of a mixture of crystalline components using X-ray diffraction techniques is based on the fact that one component produces a characteristic diffraction pattern, whether it is on its own or mixed with another. The intensity of the maximum diffraction of a certain component in the mixture depends on the concentration of that component. However, the relationship between intensity and concentration is not linear, because the diffraction intensity depends significantly on the absorption coefficient of the mixture, which, in turn, varies with the concentration of the component. The intensity of the diffracted beam takes the form:

$$I = \left(\frac{I_0 A\lambda^3}{32\pi r}\right)\left[\left(\frac{w_0}{4\pi}\right)^2 \frac{e^4}{m^2}\right]\frac{1}{v^2}\left[F^2 p\left(\frac{1+\cos^2 2\theta}{\sin^2\theta\cos\theta}\right)\right]\frac{e^{-2M}}{2w} \tag{2.3}$$

In eq. (2.3), I is the integral intensity of maximum diffraction; I_0 represents the intensity of the incident beam; A represents the cross-section of the incident beam; λ is the wavelength of the incident radiation; r is the radius of the diffractometer circle; w_0 is a constant; e represents the charge of the electron; m is the mass of the electron; v is the volume of the elementary cell; F represents the structural factor; p is the multiplicity factor; θ is the Bragg angle; e^{-2M} represents the temperature factor; and w is the linear absorption coefficient.

In order to perform a quantitative analysis by X-ray diffraction, the phases present in the sample are first identified, starting with qualitative analysis. This identification process may not be simple due to the multitude of components that may be present in a multiphase sample.

We must emphasize that, in general, a single method is not sufficient to determine the composition of a multiphase mixture. In addition to X-ray diffraction analysis, at least one elemental analysis would also be required. In the absence of this, comparison of the X-ray spectra obtained by the powder method to spectra from existing databases can be used (PDF = Powder Diffraction Files; ASTM = American Society for Testing Materials; JCPDS = Joint Committee on Powder Diffraction Standards; Pearson Database, etc.). In these databases, the values *d* and θ are in the form of tables together with their relative intensities ("intensity" expressed as a percentage of the most intense line) and Miller indices *hkl* of these lines, for over 65,000 crystalline compounds and phases.

The Rietveld method is a computerized method of calculating the diffraction figures by fitting the smallest square of a given structural model and comparing it with the diffraction pattern experimentally obtained from the sample. Rietveld first developed this method in 1969 to study the crystalline structures of powders through neutron diffraction data [24]. This method was then adapted by Malmros and Thomas (1977) [25] and Khattak and Cox (1977) [26] for data obtained from X-ray diffraction.

X-ray diffraction analysis was performed in the Laboratory for Solid Body Crystallographic Determination (LDCCS) in the Faculty of Physics at the West University of Timişoara. Phase analysis of the investigated samples was performed based on the X-ray diffraction spectrum recorded using a Bruker-AXS D8 ADVANCE diffractometer in a θ: 2θ assembly using MoK_α radiation (50 kV, 40 mA, Zr filter, and λ = 0.07093 nm). Using specialized calculation programs, the experimentally obtained diffraction maxima were indexed and the network constant determined. The net advantage of the equipment setup used in the laboratory was the reduced purchase time and the ability to analyse a small amount of the sample.

In our research, we analyzed the structural and morphological characteristics of several samples in the form of metal oxide powders, obtained by different synthesis methods. Experimental investigations were performed on the structural transformations induced by heat treatment in manganese ferrites synthesized by ultrasonically-assisted chemical co-precipitation [13], or in manganese ferrite-polyaniline hybrid materials [28] synthesized by the polymerization of aniline on the surface of $MnFe_2O_4$ using ammonium persulfate as an oxidant. The physical and structural properties of some perovskitic materials synthesized by the sol-gel method were also analyzed, including $NaTaO_3$ [12, 27], doped and undoped by metal ions of Ag [12] or Cu [27], and $LaMnO_3$ [14, 15, 29, 30], doped and undoped [14] with Ag [15], Ca [29], or Pd [30] ions.

Recently, we studied the structural and morphological properties of some oxide compounds—Cu-Mn [17] and Fe-Pd—by XRD analysis [18].

For example, for the preparation by co-precipitation of ferrite powders of $MnFe_2O_4$, in [31] the authors used iron and manganese sulfides as reactants. To obtain similar samples of $MnFe_2O_4$ by the same method, we used other reagents: manganese and iron chlorides ($MnCl_2 \cdot 4H_2O$ and $FeCl_3 \cdot 6H_2O$). We thus obtained samples of $MnFe_2O_4$, which were then calcined at different temperatures. At the same time, using the same precursors in the same amounts, a manganese ferrite sample obtained by the hydrothermal method was synthesized. Figure 2.2 shows the XRD spectra of two samples: a) sample 1 obtained by co-precipitation followed by calcination at 900 °C and b) sample 2 prepared by the hydrothermal method.

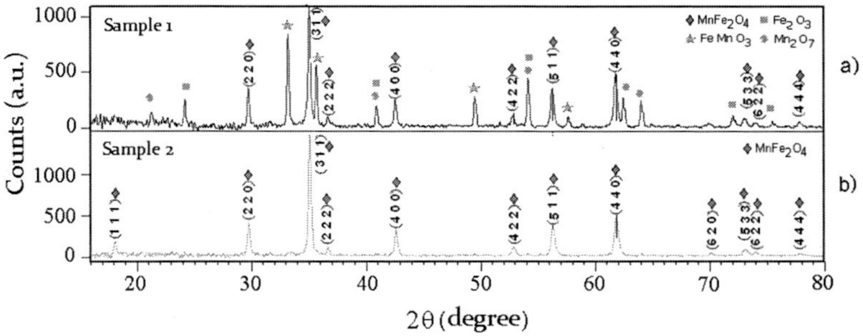

Fig. 2-2. XRD spectrum of: a) sample 1 and b) sample 2.

As can be seen from fig. 2-2 a), sample 1 consists of a mixture of $MnFe_2O_4$, $FeMnO_3$, Mn_2O_7, and Fe_2O_3-type phases, in accordance with reference JCPDS no. 75-0034 [32]. As reported in [33], at high temperatures, the ferrite $MnFe_2O_4$ is unstable in air because the Mn^{2+} ions oxidize into Mn^{3+} ions, resulting in the dissociation of the spinel-type $MnFe_2O_4$ and the production of an $FeMnO_3$ perovskite-type structure. Therefore, no method of preparation involving high temperatures is suitable for the preparation of spinel-type manganese ferrite nanoparticles. In addition, as can be observed in fig. 2.2 b), sample 2, obtained by the hydrothermal method, shows an exclusively pure phase of manganese ferrite ($MnFe_2O_4$) with a spinel-type cubic structure, in accordance with reference JCPDS no. 75-0034 [32]. In all cases, the crystalline lattice parameter *a* was determined and Scherrer's formula was used to calculate the mean size *D* of the crystallites [34, 35]. The values obtained are shown in table 2-1. Such results have been reported in [13, 16, 28] for similar samples based on metal oxides of the Mn-Fe type.

Ceramic materials with perovskitic structures (type ABO_3) can be obtained from nanocrystalline powders, their preparation being related to the distribution of cations between the tetrahedral and octahedral positions of the crystal lattice depending on the conditions and method of synthesis. Some of the best known materials with a perovskitic structure are those based on lead, such as $PbZrO_3$ and $PbTiO_3$ (PZT). They display piezoelectric behavior and are used as sensors and piezoelectric transducers [36]. In order to limit the use of lead-based systems, studies have been carried out in recent years to obtain alternative lead-free materials that are compatible with the latest environmental directives. Compounds based on niobium and tantalum oxides, such as alkaline tantalates and niobates, have been developed and are considered to be the most promising compounds in the list of functional ceramic materials for future technologies [37, 38]. Sodium tantalate ($NaTaO_3$) is a perovskitic oxide that is a feasible alternative to lead-based compounds [39]. Its morpho-structural, electrical, and dielectric properties have been extensively studied by our team in recent years [12, 14, 15, 27, 30, 40].

Table 2-1. Lattice parameters and the mean size *D* of the crystallites in the investigated samples as determined by X-ray diffraction.

Samples	Phases	Lattice parameters, [Å]	Size of crystallites, *D* [nm]
Sample 1	$MnFe_2O_4$	a=b=c=8.4989	37.93
	$FeMnO_3$	a=b=c=9.365	-
Sample 2	$MnFe_2O_4$	a=b=c=8.48898	31.9
Sample 3	$NaTaO_3$	a=b=5.508, c=3.882	50.2
Sample 4	$NaTaO_3$	-	39.1
	$Na_2Ta_4O_{11}$	-	26.6

In general, the method of synthesis for the preparation of most ceramic powders with a perovskitic structure is the solid state reaction [41]. Recent studies have allowed the development of other synthesis methods, such as the sol-gel method, the hydrothermal method, and the ultrasonically-assisted hydrothermal method [42, 43], which are all useful for the preparation of tantalates (e.g. $NaTaO_3$ type).

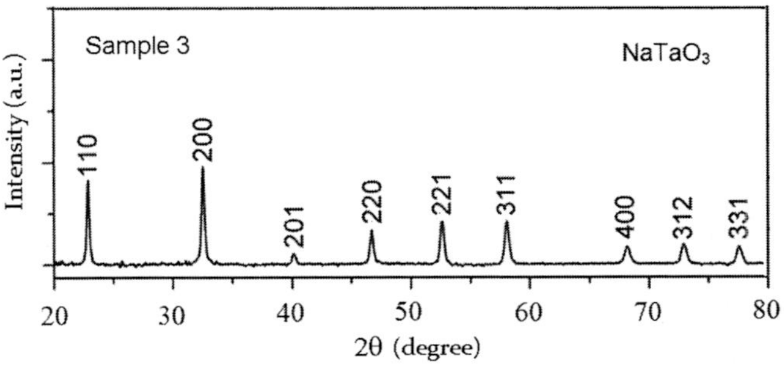

Fig. 2-3. XRD pattern for $NaTaO_3$ sample obtained by sol-gel method.

Figure 2-3 shows the XRD spectrum of a $NaTaO_3$ sample (denoted as sample 3). This sample was obtained by the sol-gel method using a solution of sodium hydroxide in ethylic alcohol and distilled water to which tantalum (V) ethoxide was added at room temperature, similar to the process described in [12, 27]. The resulting suspension was heated at 90 °C for 2 hours then the obtained gel was dried in a furnace for 4 hours at 110 °C to give a sodium tantalate powder. Figure 2-4 shows the XRD spectrum of a $NaTaO_3$ sample doped with Al ions (denoted as sample 4), obtained by the ultrasonically-assisted hydrothermal method and using tantalum ethoxide ($Ta(OC_2H_5)_5$), sodium hydroxide (NaOH), and aluminum nitrate ($Al(NO_2)_3$) as raw materials. The suspension obtained was subjected to an ultrasonic process. The resultant gel was washed with water and ethanol to remove byproducts and then dried at 80 °C for 2 hours. Finally, the powder was heat treated at 400 °C for 6 hours.

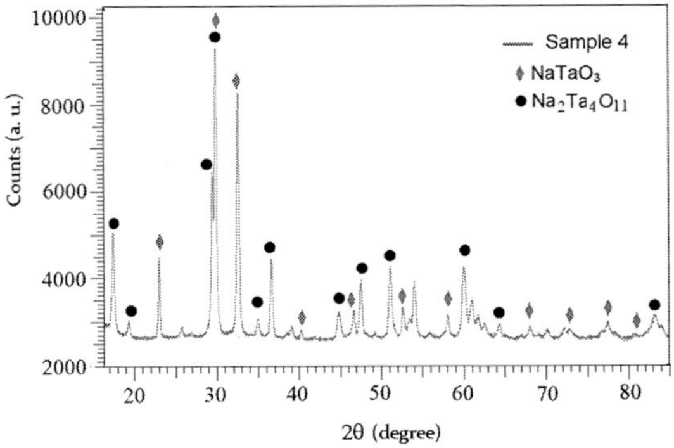

Fig. 2-4. RX spectrum of $NaTaO_3$ sample doped with Al ions obtained by the ultrasonically-assisted hydrothermal method (sample 4).

As can be seen in fig. 2-3, sample 3, obtained by the sol-gel method, was well crystallized with a pure phase of sodium tantalate ($NaTaO_3$) and a tetragonal structure, in accordance with reference JCPDS No. 01-073-0881, space group P4/m b m, and with the lattice parameters, a, b, and c listed in table 1. The mean size D of the crystallites [34, 35] was determined using Scherrer's formula and the value obtained is shown in table 2.1. Similar results have been reported in our previous papers [12, 27, 29, 30, 40] for other samples with perovskitic structures. Also, in fig. 2-4, the XRD spectrum obtained by the ultrasonically-assisted hydrothermal method (sample 4) shows low crystallinity, with the sample being in two phases: $NaTaO_3$ and $Na_2Ta_4O_{11}$ (natrotantite, according to JCPDS 38-0463). According to the X-ray diffraction data, the size of the crystallites was estimated using the Scherrer equation, the values of the average diameters of the crystallites being given in table 2.1. The quality of the diffractograms for sample 4 does not allow the calculation of the network parameters.

2.3. Scanning Electron Microscopy/Spectroscopy Energy-Dispersive X-Ray (SEM/EDX)

In the study of solid states, most analytical techniques use electron beams as probes to investigate the material. At the interaction between the accelerated electron beam and the surface of the solid material, secondary emission processes, the excitation of atoms of the material with characteristic X-ray emissions and the emission of photons or Auger electrons, can occur [44]. It is very important to know the interaction between the electrons and the substance in order to interpret the various signals obtained. Scanning electron microscopy is used for the study of surface morphology with the help of secondary or reflected electrons, affording us the possibility of examining samples with irregular surfaces and variable thicknesses of a few centimeters and height of 1 cm. The image formed is three-dimensional. At the usual acceleration voltages in electron microscopy, wavelengths are obtained that are approximately 10^4-10^5 shorter times than the wavelength of light. Thus, the resolution of electron microscopes is clearly superior to optical ones, reaching the order of some Angstroms for the most advanced instruments. The wavelength associated with the electrons accelerated below a potential difference V is given by the relation [44]:

$$\lambda = \frac{h}{\sqrt{2emV\left(1 + eV / 2mc^2\right)}} \tag{2.4}$$

where h is the Planck constant; m is the mass of the electron; e is the charge of the electron; c is the speed of light in free space; and V is the acceleration voltage.

To obtain a very narrow electron beam (less than 100 Å in diameter), the beam produced by an electronic cannon is reduced by means of 2 or 3 electromagnetic lenses. Inside the final electromagnetic lens are two deflection coils activated by a scanning current. In this way, the primary electron beam is focused and thus performs a zig-zag motion over the investigated sample, "sweeping" its surface. Surface scanning is performed either by mechanical movement of the sample in a fixed electron beam, or by deflection of the electron beam using variable electrostatic or electromagnetic fields applied in two mutually perpendicular directions [45].

Scanning electron microscopy (SEM) is valuable to the study of sample morphology because it provides information about the shape, size, and distribution of crystallites; the compactness of the deposited layer; and its thickness by sections. The basic principle of SEM operation is to apply a voltage between a conductive sample and a filament, which leads to the emission of electrons from the filament to the sample being analyzed. Measurements are made in a vacuum enclosure of 10^{-4}-10^{-10} Torr.

The electrons are oriented to the sample by means of magnetic lenses. The resolution of the obtained images depends on the electron beam current and the final size of the electron spot that can be adjusted with one or more condensing lenses [44]. The electron beam interacts with the sample in the range of a few nanometers to a few microns, depending on the parameters of the beam and the type of sample. The signal given by the secondary electrons is commonly used in morphological investigations. Secondary electrons are produced as a result of the interaction between the electron beam and the weakly bound electrons in the conduction band of the studied material. Some of the beam's energy is transferred to the electrons in the conduction band, giving them enough energy to come to the surface of the material as secondary electrons. The secondary electrons have low energies (< 50 eV) and thus only those at a very shallow depth, i.e. a few nanometers, will have enough energy to reach the surface and be picked up by the detector. High-energy electrons that have been backscattered can also form secondary electrons. As such, the SEM image obtained is the result of the intensity of the emission of secondary electrons in the sample.

Energy dispersive X-ray spectroscopy (EDAX), also called EDS or EDX, is an X-ray technique used to identify the elemental composition of materials. EDAX detectors are modules attached to scanning electron microscopes (SEM). EDAX spectra show peaks corresponding to the elements that make up the actual composition of the analyzed sample. It is also possible to determine the elemental composition at points of interest observed on the surface of the sample. EDAX is a valuable tool for qualitative and quantitative elemental analysis. Each element displays particular characteristics corresponding to the possible transitions of electrons between the various electronic layers of the electron shell (fig. 2-5) [46]. The stages of a general EDAX spectral analysis are: acquisition of the spectrum; correction for detector effects; identification of the elements present and the selection of a series of lines for each element; calculation of the environmental braking radiation; evaluation of the overlapping peaks and estimation of net intensities; calculation of concentrations by comparison or otherwise to standards; and, finally, processing and presenting the results. EDAX elemental analysis allows the quantitative investigation of the studied materials, providing information about the component elements of the material, as well as quantifying these elements in atomic and mass percentages.

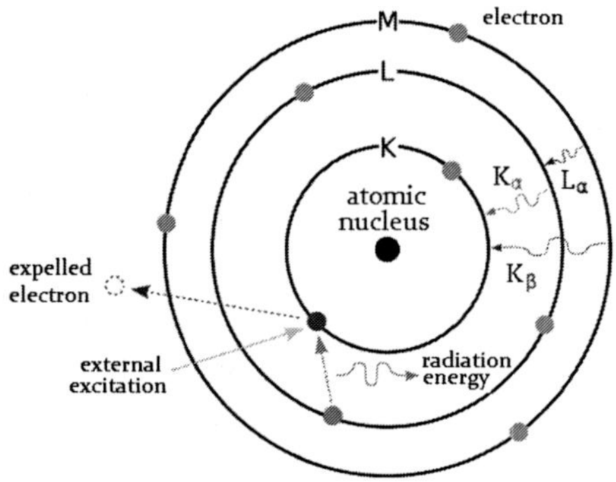

Fig. 2-5. The principle of EDAX analysis.

Characterization of the obtained samples was performed using the scanning electron microscope Inspect S (FEI Company) and Edax Inspect S, with a resolution of 133 eV, from the National Research-Development Institute for Electrochemistry and Condensed Matter (NRDIECM)

Timisoara, Department of Condensed Matter. The scanning electron microscope was used for qualitative (imaging) and quantitative (EDAX) analysis of conductive samples and/or samples prepared by conventional (coated) methods.

a)

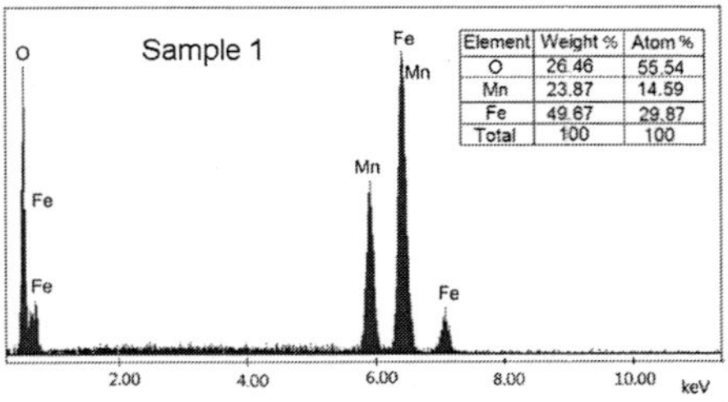

b)

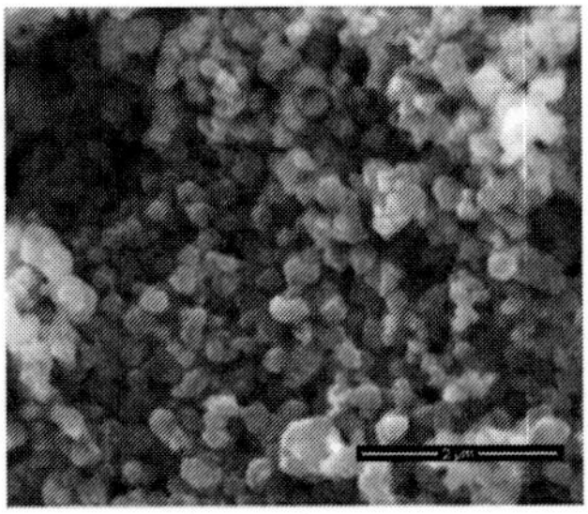

c)

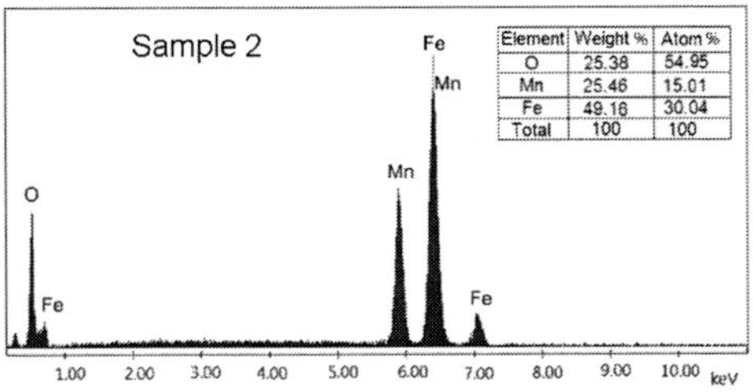

d)

Fig. 2-6. SEM images of samples 1 (a) and 2 (c); and EDX elemental analysis of samples 1 (b) and 2 (d).

Figure 2-6 a) and c) gives SEM images of samples 1 and 2 of Mn-Fe type, which were obtained by co-precipitation (sample 1) and the hydrothermal method (sample 2). From these SEM images, we can observe that the synthesized samples consist of particle aggregates of various sizes. The semi-quantitative analysis EDX of these samples is shown in fig. 2-6 b) and d), confirming that samples 1 and 2 consist of Mn, Fe, and O. The atomic ratio of the elements is 1: 2.04: 3.81 (for sample 1) and 1: 2.00: 3.66 (for sample 2), being close to the $MnFe_2O_4$ ratio.

Figure 2-7 a) gives the SEM images and element map of sample 4 obtained by the ultrasonically-assisted hydrothermal method to give $NaTaO_3$ doped with aluminum ions. In fig. 2-7 a), we can observe that, in sample 4, the crystalline cubic phase of $NaTaO_3$ is somewhat present, being formed of granules linked together in agglomerated groups. At the same time, in fig. 2-7 a) we can see the chemical composition of $NaTaO_3$ powder, which consists of O, Na, Al, and Ta, and the mapped elements attest to the homogeneity of the studied compound. The results obtained regarding the elemental analysis by EDX electron microscopy shown in fig. 2-7 b), confirm the presence of the main elements of Na, O, Al, and Ta in the sample.

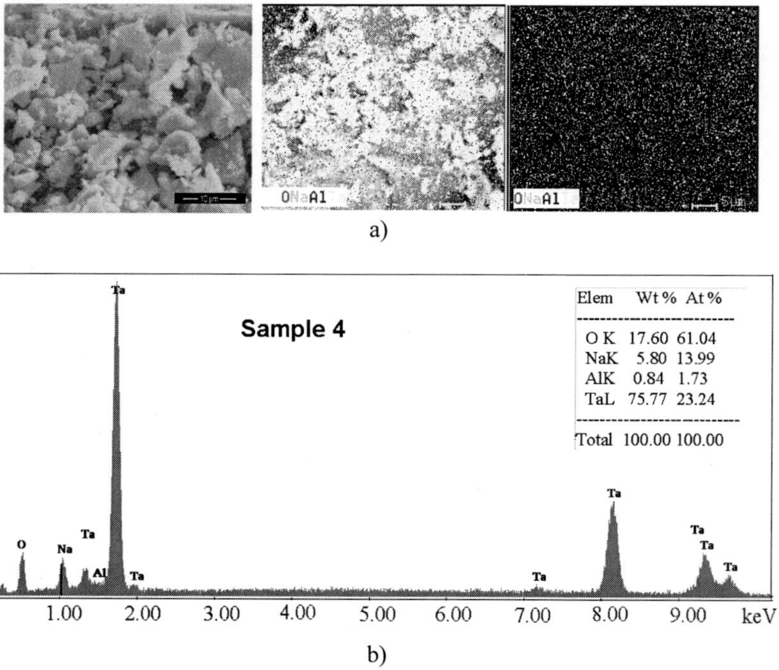

Fig. 2-7. a) SEM images and map of elements of an $NaTaO_3$ sample doped with Al ions (sample 4); b) EDX/quantification of sample 4.

The morpho-structural properties of some similar samples based on metal oxides of Mn-Fe, Na-Ta, Fe-Cu, and Fe-P were investigated and the results obtained by our team [12-18, 27-30, 40] were published.

2.4. UV-VIS Spectroscopy

Visible and ultraviolet spectroscopy (UV-VIS) is an analytical technique for investigating matter. Spectroscopy measures the absorption, transmission, scattering, or emission of electromagnetic radiation through interactions with atoms or molecules. Absorption represents the transfer of electromagnetic energy from the source to an atom or molecule. In scattering, light is redirected after interacting with the matter being studied and the emission is generated by the electromagnetic energy changing from one energy level to another. This technique is focused on measuring the interactions of radiant energy with matter, using appropriate equipment. UV-VIS spectroscopy is an optical method of material analysis that uses electromagnetic radiation to excite the component atoms and molecules. It is used to obtain information on the nature and quantities of a material's constituents. An electromagnetic radiation spectrum is obtained by decomposing the radiation using a spectroscope and consists of a succession of images of the input slit formed by the different monochrome radiations of the incident light [47].

Such spectra can be classified as follows: a) emission spectra obtained from sample particles in excited states (by thermal excitation) with higher energy, before returning to their base state with lower energy, [48]; b) absorption spectra obtained by irradiating a sample with radiation whose wavelength (or frequency) continuously varies so that the sample particles transition into excited states with certain characteristic parts absorbed [48].

Samples analyzed using UV-VIS spectrophotometry can include: solid materials; liquid samples; crystals; and metallic plates. The analysis of solid samples is done with the diffuse reflectance module, while liquid ones are analyzed with the absorption/transmission module. UV-VIS spectroscopy is used as an analytical method for two reasons. The first is that it can be used to identify functional groups in molecules and the second is that it can be used to verify these functional groups.

The electronic transitions that occur when a molecule changes from its fundamental state to an excited state are associated with certain types of spectra. The interpretation of the resultant spectrum can provide information about molecular and atomic energy levels, the distribution of species between these levels, molecular geometry, chemical bonding, and the interactions of molecules in solution. At the same time, by comparing the spectra, qualitative basic information can be obtained regarding the composition and chemical structure [49].

UV-VIS spectroscopic analysis of the samples was performed using the Lambda Perkin-Elmer 950 DRUV-VIS spectrometer at the National Research-Development Institute for Electrochemistry and Condensed Matter (NRDIECM), Timisoara, which works in the wavelength range 250-600 nm. Figure 2-8 shows the UV-VIS absorption spectrum for an $NaTaO_3$ sample doped with Al ions obtained by the ultrasonically-assisted hydrothermal method (sample 4, see paragraph 2.2). The wavelength λ_g, which is at the end of the absorption range, was determined as the intersection between the wavelength axis and the tangent line drawn on the absorption spectrum [50].

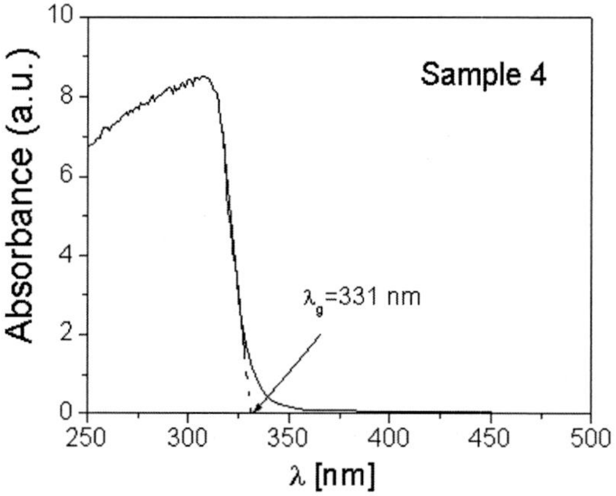

a)

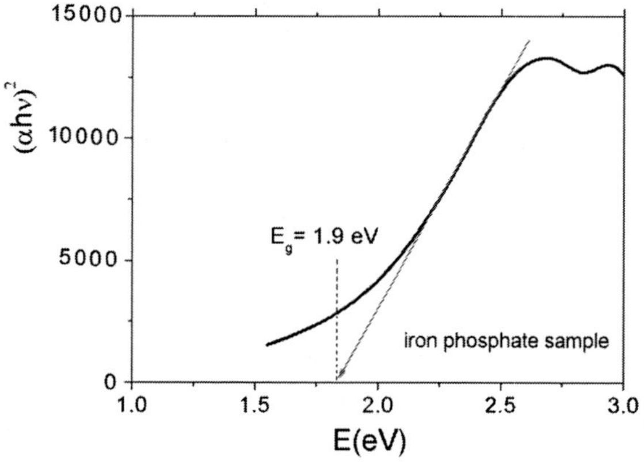

b)

Fig. 2-8. UV-vis spectrum for: a) NaTaO3 sample doped with Al ions (sample 4); b) iron phosphate sample ($Fe_3(PO_4)_2 \cdot 8(H_2O)$).

As can be seen in fig. 2-8 a), the value of the wavelength λ_g for sample 4 is $\lambda_g = 331nm$. The value of the band gap energy E_g can be computed from the UV-vis spectrum using the following equation [51]:

$$E_g(eV) = \frac{1240}{\lambda_g(nm)} \tag{2.5}$$

Equation (2.5) and the value found experimentally for λ_g in fig. 2-8 a) gives the following value for the band gap energy E_g = 3.74 eV (for sample 4), which is close to the values (4.0-4.1 eV) given in other work [52, 53] for undoped samples of $NaTaO_3$ obtained by the sol-gel method, or the value of 3.8 eV for samples of $NaTaO_3$ obtained by the ultrasonically-assisted hydrothermal method [54, 55].

A sample of iron phosphate of the vivianite type with the chemical formula $Fe_3(PO_4)_2 \cdot 8(H_2O)$ obtained by the co-precipitation method was analyzed [56]. Figure 2-8 b) shows the UV-VIS spectrum of this sample, i.e. the dependence $(\alpha E_{photon})^2$ on E_{photon} energy for a permitted direct transition, where α is the absorption coefficient and $E_{photon} = h\nu$ is the photon energy (where h is the Planck constant and ν is the frequency). The value of E_{photon} extrapolated to $\alpha = 0$ gives an absorption energy that corresponds to the optical band gap energy E_g. In figure 2.8 b) we can observe that the value obtained is $E_g = 1.9\ eV$. Other papers [57, 58] indicate values close to the value obtained by us for E_g, corresponding to samples of the same type that contain iron and phosphorus ions. For example, in [58] the authors show that the optical properties of samples doped with phosphorus and iron ions are strongly influenced by the increase in the concentration of iron ions. Thus, with variation of the concentration from 0.5 to 1 %, the energy of the optical band gap E_g decreased from 2.62 eV to 1.95 eV [58] and with a value close to the value obtained by us for the vivianite sample (fig. 2-8 b)).

2.5. Fourier Transform Infrared Spectroscopy (FTIR)

Fourier transform infrared spectroscopy (FTIR) is a technique used to obtain an absorption or emission infrared spectrum for a solid, liquid, or gaseous material. The name *Fourier transform infrared spectroscopy* arises from the fact that to convert the data generated into the real spectrum a Fourier transform is required, i.e. the application of mathematical processing [59].

FT-IR is an analytical method that relies on the absorption of electromagnetic radiation in the infrared (IR) range by some materials. With the interaction of condensed matter and electromagnetic radiation in the infrared spectral domain, 2.5-25 μm (4000-400 cm^{-1}) excites vibrational movement in atoms (ions) and molecules, deforming the bonds between them. Two characteristic types of vibration are highlighted in the form of absorption bands, namely: 1) "stretching vibrations" (along the bonds); and 2) "bending vibrations" (modification of the angles between the bonds). In the near IR domain, the obtained absorption bands allow the identification of these bonds, as well as the quantification of different functional groups (C = O, O-H, C = S, CH3, CH2, and C-NO2, etc.). Based on a large number of recorded spectra (several thousand), Colthup et al. developed graphs that correlated the vibration frequencies with the nature of the vibrating atoms within the bond for a significant number of groups of atoms [60]. Each spectral maximum is associated with the vibration of a certain bond in a molecule of the sample. The function of such correlation graphs is to suggest possible compounds and structures for further investigation.

FT-IR spectra make it possible to identify all types of organic and inorganic compounds, determine the functional groups in organic materials, identify the chromatographic elements present, quantitatively determine compounds in mixtures, determine the composition of molecular surfaces, and determine the molecular structure (isomers) and stereochemistry of molecules. Vibrational spectroscopy was developed on this basis and has become a common method of material analysis for samples in all three states of aggregation. The basic equipment includes FTIR spectrophotometers (spectrometers with signal/noise ratio improved by filtering, using the Fourier Transform method, of a large number of successively recorded spectra).

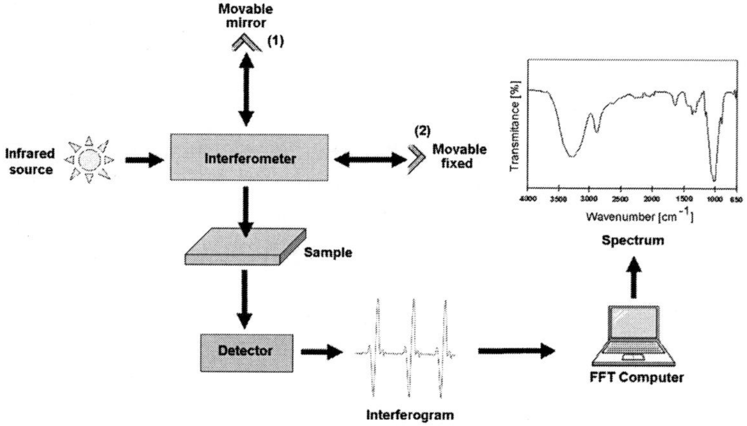

Fig. 2-9. A spectrometer based on Fourier transform infrared spectroscopy.

Figure 2-9 shows the basic scheme of a Fourier transform infrared (FTIR) spectrometer. These spectrometers are more advanced and accurate than dispersive spectrometers [61]. Instead of the diffraction grating of dispersive spectrometers, the FTIR spectrometer uses a Michelson interferometer to generate the dispersed wavelengths. Thus, in the Michelson interferometer, in (1) the mirror moves continuously with a constant velocity, while in (2) the mirror is fixed (fig. 2-9). As a result, the distance travelled by the moving mirror beam (1) is different from the distance travelled by the fixed mirror beam (2). In the interferometer, the recombination of the radiation beams with different distances travelled takes place, leading to the creation of an interference pattern called an interferogram, which can be used to obtain a single beam spectrum by Fourier transform. Furthermore, the beam interacts with the sample, which will absorb the energy corresponding to the characteristic and unique frequencies in the interferogram. The infrared absorption spectrum can be obtained by comparing the spectrum with a single background beam and that of the sample [62]. Finally, the spectrum of the sample is obtained by computation of the Fourier transform using software and the fast Fourier transform algorithm (FFT) [59].

For the examination of metal oxide samples prepared by different methods [28], the Shimadzu IR Prestige-21 spectrometer at the Faculty of Chemical Engineering, University Politehnica of Timişoara (UPT) was used, having the spectral domain 240-12,500 cm^{-1}; a spectral resolution of 0.5-2 cm^{-1} (in close IR); and a signal-to-noise ratio of 92.5 dB.

The spectrometric samples were prepared by incorporating powders synthesized in KBr tablets under pressure. This was done to avoid distortions of the absorption bands by the Christiansen effect [64, 65], which occurs due to a possible anomalous dispersion (refractive index) in the absorption band, as well as distortion of the spectra due to the uneven thickness of the support tablets (the effect manifested by widening of the absorption bands, as well as increasing the intensity of the weak absorption bands, both unjustified). In addition, we took into account the fact (undesirable) that both too extensive sample granulation and insufficient homogenization in the KBr pastille produce artefacts, thus altering the real absorption spectrum. Once these precautions were taken, four samples based on metal oxides were analyzed by the Fourier transform infrared spectroscopy (FTIR) technique, as follows:

Two samples of ferrite-type Mn-Fe (samples 1 and 2, see table 2.1) were obtained by co-precipitation followed by calcination at 900 °C (sample 1) and by the hydrothermal method (sample 2).

Two $NaTaO_3$ type samples were obtained: one undoped and obtained by the hydrothermal method, denoted as the SH sample, and the second $NaTaO_3$ sample doped with Cu ions obtained by the ultrasonically-assisted hydrothermal method, denoted as the US-Cu sample.

The IR absorption spectra were recorded under identical conditions for the investigated samples, as presented in figs. 2.10 and 2.11 a), b).

Data in the existing literature were used to identify the relevant absorption bands [66]. Figure 2-10 shows that the FT-IR spectrum of samples 1 and 2 (see table 2-1) confirms the presence of $MnFe_2O_4$ in sample 2, highlighting the characteristic bands of this substance and the presence of multiple phases in sample 1. Sample 2 shows an intense band at 570 cm^{-1} confirming the presence in a single phase of $MnFe_2O_4$, in accordance with the XRD analysis of this sample in fig. 2-2. b).

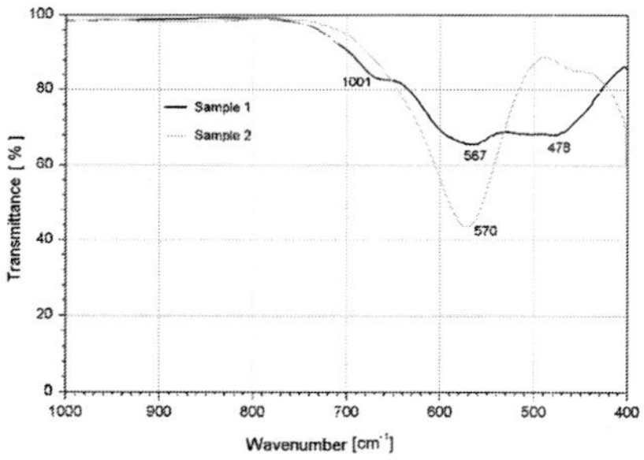

Fig. 2-10. FT-IR spectra of Mn-Fe samples.

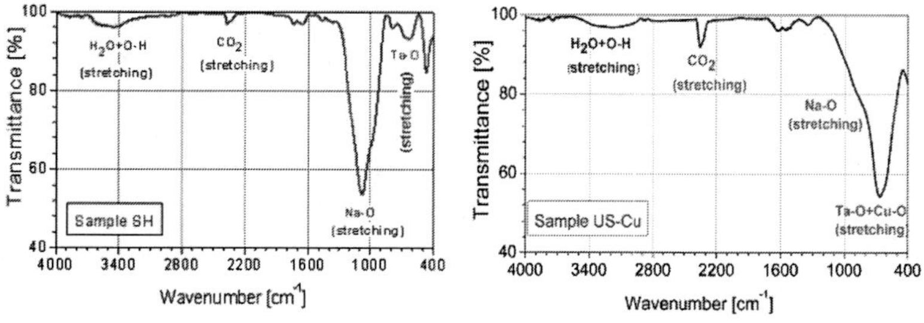

a) b)

Fig. 2-11. FT-IR spectra of sample of undoped $NaTaO_3$ SH (a) and sample $NaTaO_3$ doped with Cu ions obtained by the ultrasonic method (b).

Analyzing the results obtained in fig. 2-11 a) and b), the following can be stated: 1) the SH sample of pure $NaTaO_3$ (fig. 2-10) shows an IR absorption spectrum in which the main absorption band appears (at 1,100 cm^{-1}) attributed to vibrations (stretching) in the Na-O functional group, as well as three distinct absorption bands (weaker), which indicate the excitation of some vibrations (stretching) of the Ta-O group; 2) the presence of doping Cu ions in the structure of $NaTaO_3$ prepared by the ultrasonic method (US-Cu, fig. 2.11 b)), causes a significant decrease, up to extinction, of the share of the main absorption band (Na-O), accompanied by the appearance a strong absorption band attributed to the excitation of some vibrations (stretching) of the Ta-O and Cu-O groups, with a much higher share of the latter. It is important to note that such changes in the IR absorption spectra of $NaTaO_3$ following doping with metal ions (Cu in this case) are little represented in the literature and so studies of this type will be the subject of future research by our team.

2.6. Thermogravimetric Analysis (TGA)

Thermal analysis methods include thermogravimetric analysis (TGA) and differential thermoanalysis (DTA). TGA records the mass of the samples as a function of time or temperature (thermogravimetric curve), while the temperature of the samples is programmed in a controlled atmosphere. This analytical technique is used in determining the composition of samples and predicting their thermal stability. From the TGA analysis, information can be provided on chemical and physical phenomena, including chemisorption, dehydration, decomposition, gas-solid reactions (e.g. oxidation or reduction), vaporization, sublimation, absorption, and desorption. As a result, TGA can evaluate samples that show a decrease or increase in mass due to decomposition, oxidation, or dehydration [67]. There are several thermogravimetric (TGA) techniques: 1) static or isothermal thermogravimetry in which the weight of the sample is recorded as a function of time at a constant temperature; 2) quasi-static thermogravimetry in which the sample temperature is increased in steps, separated by isothermal intervals, so that in the time between two intervals, the sample mass reaches stability; and 3) dynamic thermogravimetry in which the sample is heated in an environment whose temperature is changed linearly.

The instrument used for thermogravimetric analysis (TGA) is called a thermogravimetric analyzer and it can continuously measure the mass of a sample as its temperature changes over time. It should be specified that, in thermogravimetric analysis, mass, time and temperature are basic measurements from which other additional measurements can be derived. A thermogravimetric analyzer consists of a precision balance with a sample pan located in an oven whose temperature can be programmably increased at a constant rate to withstand the thermal reaction. This thermal reaction can take place in a variety of atmospheres (air, vacuum, inert gas, oxidizing/reducing gases, or liquid vapor), as well as a variety of pressures (constant pressure, high pressure, or high vacuum).

For the TGA analysis of some metal oxide samples, we used the LabSys Evo instrument at I.N.C.D.E.M.C. Timisoara, which covers a temperature range from room temperature up to 1,600 °C and can work in various modes of analysis, i.e. TGA and simultaneously TG-DTA.

The structure of the LabSys Evo thermal analysis system consists of: a TG balance with TDA sensors; an oven with a metal resistor element; a gas circuit (for the selection and application of a specified atmosphere to drive the sample reaction: air, argon, and vacuum); and data acquisition software.

Figure 2-12 presents the temperature dependence of the variation in weight of a CuMnO2 sample, ranging from room temperature up to 900 °C in air, with a rate of 10 °C/min, which represents the TG-DTA curve.

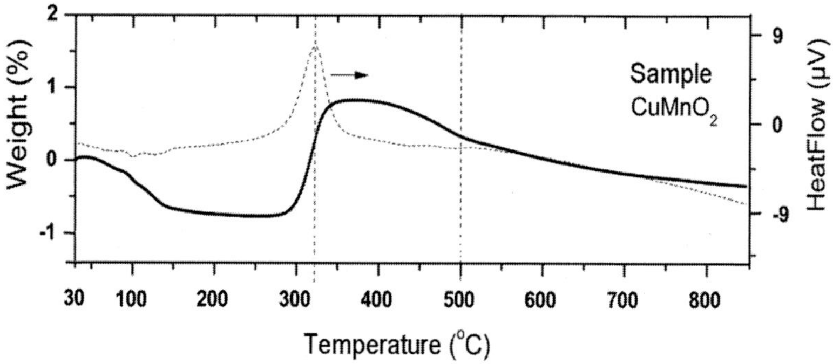

Fig. 2-12. The TG–DTA curve obtained for a $CuMnO_2$ sample: the left y axis expresses the mass variation (with a continuous line) and the right y axis expresses the heat flow (dotted line).

As can be seen in fig. 2-12, thermogravimetric analysis performed in air for a sample of $CuMnO_2$ type indicates a first loss in weight (up to 150 °C), which can be attributed to the loss of crystallization water. Increasing the temperature, we can observe an increase in weight at about 300 °C, which can be correlated with the oxidative decomposition of the $CuMnO_2$

sample in a spinel phase ($CuMn_2O_4$) and CuO oxide, as has been shown in a number of papers [68].

2.7. Complex Impedance Spectroscopy

In order to analyze the electrical properties of a variety of different materials (ionic, semiconducting, or even insulating,) both in powder and solid state form, the technique of complex impedance spectroscopy (CIS) is used [69] for a wide range of frequencies (20 Hz-2 MHz) and temperatures between 25 and 200 °C. Using this method, information about the electrical conductivity or the dielectric constant can be obtained, based on complex impedance measurements depending on the frequency and temperature.

The electrical properties of a material can be represented in terms of complex impedance $Z^* = Z' - iZ''$, complex electrical modulus, $M^* = M' - iM''$, and complex dielectric permittivity $\varepsilon^* = \varepsilon' - i\varepsilon''$, which are connected by the relations:

$$M^*(\omega) = \frac{1}{\varepsilon^*(\omega)} = i(\omega C_0)Z^*(\omega) = M' + iM'' \tag{2.6}$$

In this equation, (Z′, M′, ε′) and (Z″, M″, ε″) represent the real and imaginary components of impedance, electric modulus, and dielectric permittivity, respectively; $\omega = 2\pi f$ is the angular frequency; $i = \sqrt{-1}$ is the imaginary factor; and C_0 is the capacity in the absence of material between the capacitor plates.

Impedance analysis in an alternating field is a modern method of studying the electrical properties of samples placed in a capacitive cell connected to the terminals of a variable frequency RLC-meter. The equivalent electrical circuit of the cell is shown by the diagram in fig. 2-13 a), where *C* and *R* represent the parallel elements of the material between the capacitor plates and R_1 represents the ohmic resistance due to the electrodes, which is sometimes neglected.

According to the circuit scheme in fig. 2-13 a), the impedance *Z* is given by the relation:

$$Z = R_1 + \frac{R}{1 - i\omega CR} = R_1 + \frac{R}{1 + \omega^2 C^2 R^2} + i\frac{\omega CR^2}{1 + \omega^2 C^2 R^2} \tag{2.7}$$

This gives the real component (Z′) and the imaginary component (Z″) of complex impedance:

$$Z' = R_1 + \frac{R}{1+\omega^2 C^2 R^2} \qquad Z'' = \frac{\omega C R^2}{1+\omega^2 C^2 R^2} \tag{2.8}$$

Measurement of the complex impedance of a sample provides useful information about the real component Z' (or resistance R) and the imaginary component Z'' (or capacitive reactance $-X$) of the complex impedance. In the CIS technique, the imaginary component Z'' of the complex impedance is plotted according to the real component Z', thus obtaining a Nyquist diagram [70] (fig. 2-13 b).

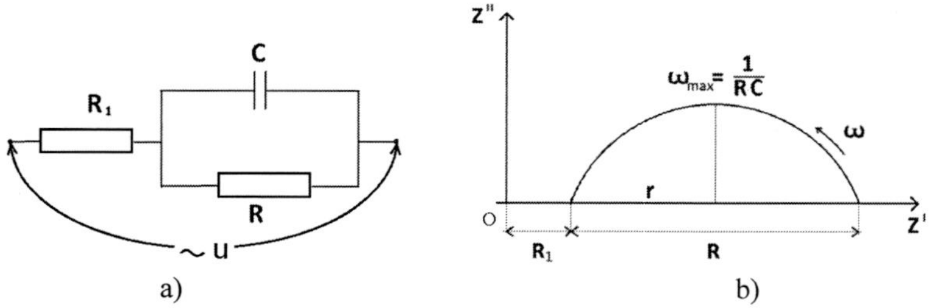

Fig. 2-13. a) Equivalent circuit of the capacitive cell used to measure complex impedance; b) Nyquist plot Z″(Z′) of complex impedance.

By removing the frequency ω from equation (2.8), we get the following:

$$\left(Z' - R_1 - \frac{R}{2}\right)^2 + Z''^2 = \left(\frac{R}{2}\right)^2 \tag{2.9}$$

As can be seen, relation (2.9) respects the equation of the circle in the $Z'OZ''$ plane with its center $C(R_1 + R/2;\ 0)$ and radius $r = R/2$ (fig. 2.13 b)). The resistance R is determined from the diameter of the semicircle and the capacitance C is calculated from the frequency corresponding to the maximum value of the semicircle.

From a microstructural point of view, the model of a composite sample is assumed to be composed of conductive plates in parallel (attributed to grains/particles), separated by resistive plates (corresponding to grain boundaries) [71, 72]. For such materials, two circular arches are observed. The first semicircle, from low frequencies, represents the grain boundaries, and the second semicircle, recorded at high frequencies, corresponds to the properties of the material in the form of grains/particles.

2.7.1. Experimental installation in a laboratory for measuring the frequency and temperature dependence of complex impedance

In order to measure the real component Z' (or resistance R) and the imaginary component Z'' (or capacitive reactance $-X$) of the complex impedance Z^* of the samples in both powder and solid form and at different temperatures T in the range 30-200 °C, we designed and realized an experimental laboratory setup, the scheme of which is given in figure 2-14.

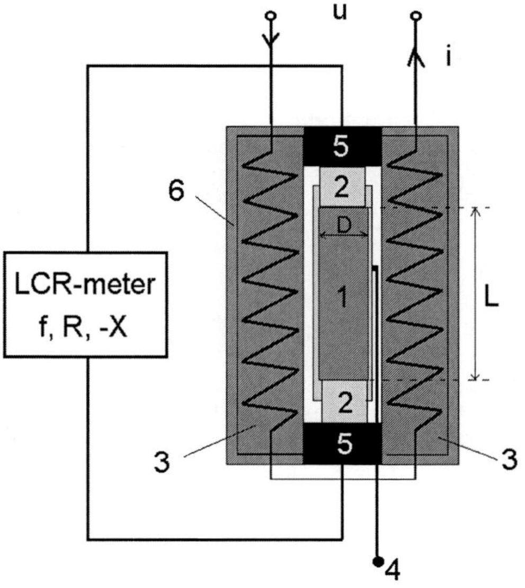

Fig. 2-14. Experimental setup for measuring the temperature dependence of the real (resistance R) and imaginary (reactance $-X$) components of the complex impedance of samples.

In fig. 2-14, we can see that the investigated sample (1), with length L and cross-section S in a glass tube of diameter D, is placed in an electric oven (3). This is then heated to temperature T with the help of an electrical resistor connected to a voltage source u. The sample (1) is in contact with two electrodes (2), which are connected to an LCR-meter of variable frequency, f, representing the capacitive measuring cell. The ends of the oven are insulated with two plugs of thermal insulation material (5) and the temperature T of the sample can be measured with a thermocouple (4). The installation is thermally insulated from the outside by inserting it into a thermal insulation shell (6). The length L of the sample varies depending on the amount of powder or the size of the solid sample obtained by pressing.

2.7.2. Determination of the components of complex dielectric permittivity

For a constant temperature T and with the sample between the plates of the capacitive measuring cell connected to the LCR-meter (fig. 2-14), the component Z′ (resistance R) and the component Z″ (reactance $-X$) of the complex impedance of the sample are measured at each frequency f in the investigated range. Based on the experimental values obtained and knowing the length L and the cross-section S of the sample, the complex resistivity $\rho^* = \rho' - i\rho''$ can be determined, using the well-known relation $R = \rho L/S$, which can be written in a complex form as $Z^* = \rho^* L/S$. As a result, the real ρ' and imaginary ρ'' components of complex resistivity can be written as:

$$\rho' = \frac{Z'S}{L} \quad , \quad \rho'' = \frac{Z''S}{L} \tag{2.10}$$

It is known that the inverse of the complex electrical resistivity represents the complex electrical conductivity σ* of a sample:

$$\sigma^* = \frac{1}{\rho^*} = \frac{1}{\rho' - i\rho''} = \sigma' - i\sigma'' \tag{2.11}$$

where σ′ and σ″ represent the real and imaginary components of the complex electrical conductivity:

$$\sigma' = \frac{\rho'}{\rho'^2 + \rho''^2} \quad , \quad \sigma'' = \frac{\rho''}{\rho'^2 + \rho''^2} \tag{2.12}$$

Taking into account relation (2.10), relation (2.12) becomes:

$$\sigma' = \frac{Z'}{\left|Z^*\right|^2}\frac{L}{S}$$

$$\sigma'' = \frac{Z''}{\left|Z^*\right|^2}\frac{L}{S} \qquad (2.13)$$

where $\left|Z^*\right| = \sqrt{Z'^2 + Z''^2}$ represent the modulus of the complex impedance of the sample. Therefore, relation (2.13) allows the determination of the components σ′ and σ″ of the complex electrical conductivity, based on the complex impedance measurements and the sample dimensions from the measuring cell.

The connection between the complex electrical conductivity σ* and the complex relative dielectric permittivity ε* of a sample [70] is given by the relation:

$$\sigma^* = i\omega\varepsilon_0\varepsilon^* = i\omega\varepsilon_0\left(\varepsilon' - i\varepsilon''\right) = \omega\varepsilon_0\varepsilon'' + i\omega\varepsilon_0\varepsilon' \qquad (2.14)$$

Using relations (2.13) and (2.14), the real ε' and imaginary ε'' components of the complex dielectric permittivity can be obtained:

$$\varepsilon' = \frac{1}{\omega\varepsilon_0}\cdot\frac{Z''}{\left|Z^*\right|^2}\cdot\frac{L}{S} \qquad (2.15)$$

$$\varepsilon'' = \frac{1}{\omega\varepsilon_0}\cdot\frac{Z'}{\left|Z^*\right|^2}\cdot\frac{L}{S} \qquad (2.16)$$

Another important parameter in the dielectric response of a sample is the complex electrical modulus *M** defined by relation (2.6), which results in the components *M*′ and *M*″ of the electrical modulus:

$$M' = \frac{\varepsilon'}{\varepsilon'^2 + \varepsilon''^2} \quad , \quad M'' = \frac{\varepsilon''}{\varepsilon'^2 + \varepsilon''^2} \qquad (2.17)$$

In conclusion, relations (2.15) and (2.16) show the frequency dependence of the components ε' and ε'' of the complex relative dielectric permittivity at temperature T in the investigated range. We can observe that the components ε' and ε'' can be determined if we know both the geometry of the sample (length L and cross-sectional area S) and the real Z' and imaginary Z'' components of its complex impedance. At the same time, it must be specified that ε'', determined in this way, expresses the total (or effective) dielectric losses in the material, due to both conduction losses and dielectric relaxation losses [73].

2.7.3. Experimental results

Using the experimental installation shown in fig. 2-14 and a LCR-meter Agilent E-4980 type, we determined the Z' component (resistance R) and the Z'' component (reactance $-X$) of the complex impedance for two samples of sodium tantalate ($NaTaO_3$) [55] in the frequency range 20 Hz-2 MHz at room temperature. The two samples (denoted A and B) were obtained using the hydrothermal method at different sintering temperatures: 800 °C (sample A) and 600 °C (sample B), respectively, but with the same time interval (12 hours) [27, 55].

The experimental dependence of frequency of the real (Z') and imaginary (Z'') components of the complex impedance for samples A and B, in the investigated frequency range (20 Hz-2 MHz) at room temperature, are shown in fig. 2-15 a) and b).

As shown in fig. 2-15, the real component Z' of the complex impedance decreases sharply with increasing frequency for sample B (fig. 2-15 b)), while for sample A, the decrease is slower at the beginning of the frequency range (fig. 2-15 a)). As a result, the electrical conductivity of both samples increases with increasing frequency. The imaginary component $Z''(f)$ has a single maximum for sample A, but two maxima for sample B (fig. 2-15). The two maxima obtained for sample B, one at low frequency ($f_{max,1}$) and the second at high frequency ($f_{max,2}$), correspond to the two relaxation processes due to the grains/particles and grain boundaries, respectively [74].

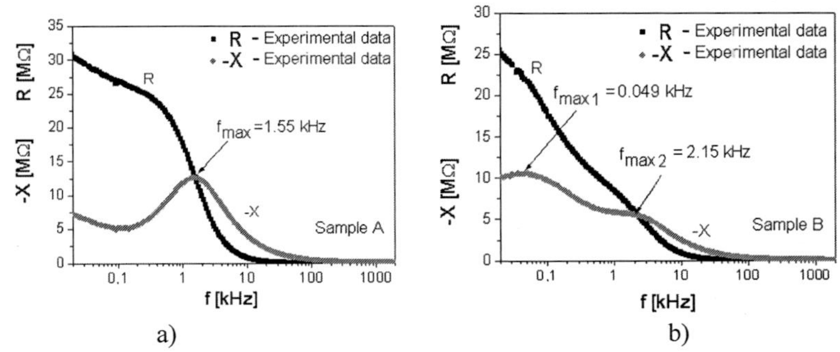

Fig. 2-15. Frequency dependence of the real component Z' (resistance R) and imaginary component Z'' (reactance $-X$) of the complex impedance for samples A (a) and B (b).

The experimental values for $f_{max,1}$ and $f_{max,2}$ for both samples are given in table 2-2. According to Debye's theory [75], the relationship between the maximum frequency f_{max}, at which the imaginary component $Z''(f)$ has a maximum, and the relaxation time τ is given by the equation:

$$2\pi f_{\max}\tau = 1 \tag{2.18}$$

Using the experimental values of the f_{max} frequency from table 2-2 in equation (2.18), we were able to calculate the relaxation time corresponding to each sample. The values obtained for the relaxation times τ are given in table 2-2.

Table 2-2. The parameters of the equivalent scheme related to the investigated samples.

Sample	$f_{max,1}$ [Hz]	$f_{max,2}$ [Hz]	τ_1 [ms]	τ_2 [μs]	R_G [MΩ]	C_G [pF]	R_{GB} [MΩ]	C_{GB} [nF]
A	-	1550	-	102.73	27.30	3.47	-	-
B	49	2150	3.38	74.06	12.20	5.60	24.10	0.14

Figure 2-16 presents the dependence of the imaginary component Z'' on the real component Z' of the complex impedance over the entire frequency range at room temperature for the investigated samples [72, 74].

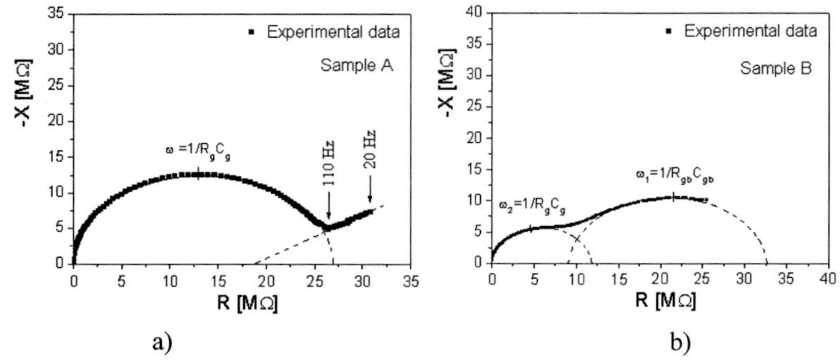

Fig. 2-16. Nyquist plot Z″(Z′) of complex impedance for sample A (a) and sample B (b).

In the case of sample A (fig. 2-16 a)), the Z″(Z′) dependence presents a semicircle with a linear portion at the beginning of the low frequency range (between 20 and 110 Hz). This behavior of the sample can be correlated with its structure, which shows that sample A has a single phase ($NaTaO_3$) [55]. As seen in fig. 2-16 b), the Z″(Z′) dependence for sample B has two semicircles, which shows that the structure of this sample has two phases ($NaTaO_3$ and Ta_2O_5) [27, 55]. The electrical responses of the samples can be represented by two equivalent parallel RC circuits connected in series (as shown in fig. 2-17) [70, 76], where the parameters R_G, C_G, and $\omega_G = 1/R_G C_G$ correspond to the grains and the parameters R_{GB}, C_{GB}, and $\omega_{GB} = 1/R_{GB} C_{GB}$ are assigned to the grain boundaries.

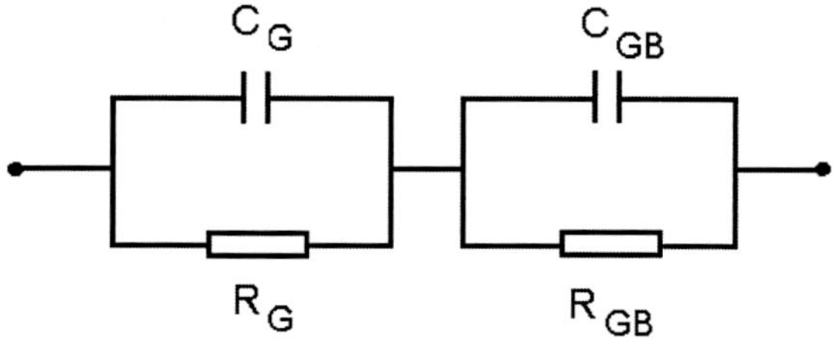

Fig. 2-17. Electrical model for the analysis of impedance spectroscopy data.

Taking into account that resistance R is determined from the diameter of the semicircle and capacitance C is calculated from the frequency corresponding to the maximum value of the semicircle, the values for the parameters corresponding to the equivalent scheme related to the investigated samples were determined. The values obtained are listed in table 2.2.

2.7.4. Thermoelectric analysis of an $MnFe_2O_4$ ferrite sample using complex impedance measurements: experimental results

Thermal analysis of the electrical properties of materials is a technique of materials science that can provide information about various phenomena (such as phase transition, adsorption/desorption, and chemical reactions), so that we can use these materials in photovoltaic, catalytic applications, and other engineering applications. In order to perform the thermoelectric analysis, we consider the manganese ferrite sample ($MnFe_2O_4$) described in the previous paragraphs, which we denoted sample 2 (see table 2-1). This sample was characterized by X-ray diffraction (XRD) (see fig. 2-2), scanning electron microscopy (SEM), and energy dispersive X-ray spectroscopy (EDX) (see fig. 2-6), and FT-IR spectroscopy (see fig. 2-10), confirming the existence of a single phase in the chemical structure of the sample.

Based on the measurements of the impedance spectroscopy over the frequency range 20 Hz-2 MHz and at different temperature values in the range 30-130 °C, a thermoelectrical analysis of the manganese ferrite sample was performed (sample 2). The low frequency region of the electrical conductivity of the sample was analyzed based on the variable-range hopping model (VRH model) [77] and the high frequency region of the electrical conductivity of the sample was interpreted according to the correlated-barrier hopping model (CBH model) [78].

Frequency dependence of the real (Z') and imaginary (Z'') components of the complex impedance $Z = Z' - iZ''$ of the investigated sample, in the frequency range 100 Hz-2 MHz and at temperatures between 30 and 130 °C, was measured using the installation shown in fig. 2-14 and an LCR-meter (Agilent type E4980A). The frequency and temperature dependence of the real (Z') and imaginary (Z'') components of the complex impedance of the manganese ferrite sample $MnFe_2O_4$ (sample 2) is shown in fig. 2.18.

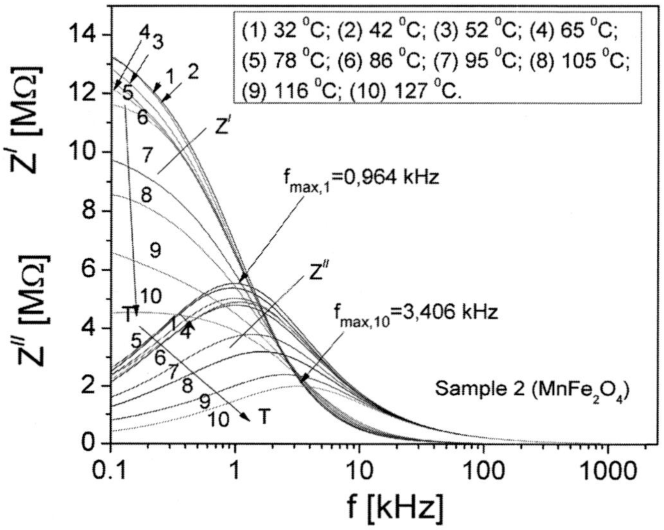

Fig. 2-18. Frequency and temperature dependence of the complex impedance components of the $MnFe_2O_4$ sample.

As can be seen in fig. 2-18, for each constant temperature *T*, the real component *Z′* of the complex impedance decreases with increasing frequency. At the same time, in the temperature range 32-60 °C (at low frequencies), *Z′* decreases with increasing temperature after which it increases slightly to 70-80 °C, and then decreases in the range 80-128 °C. The imaginary component *Z″* has a maximum at a frequency f_{max} at all temperatures investigated, as shown in fig. 2-18. This behavior suggests the existence of an electrical relaxation process in the sample [79], which is due to the presence of charge carriers [80, 81]. The maximum position of the component allows the determination of the relaxation time (τ) using the Debye relation (2.18) for each temperature investigated. The values obtained for the corresponding relaxation times at the investigated temperatures *T* are given in table 2-3.

Table 2-3. Relaxation time values as a function of temperature.

T[^{0}C]	32	42	52	65	78	86	95	105	116	128
$\tau[\mu s]$	165.18	156.91	149.58	159.44	144.40	151.22	113.00	90.85	57.93	46.74

As can be seen in table 2-3, the relaxation time τ decreases with temperature *T*, both at the beginning of the range 32-65 °C and at higher temperatures of 86-128 °C. This decrease in the relaxation time in the noted temperature range is in accordance with the Arrhenius law [80] with the form:

$$\tau = \tau_0 \exp\left(\frac{E_{A,relax}}{kT}\right) \tag{2.19}$$

where τ_0 represents the pre-exponential factor of the relaxation time τ; *k* is the Boltzmann constant; and $E_{A,relax}$ is the activation energy of the sample due to the relaxation processes. As a result, by logarithmic transformation of equation (2.19), we obtain:

$$\ln\tau = \ln\tau_0 + \frac{E_{A,relax}}{k}\cdot\frac{1}{T} \tag{2.20}$$

Activation energy $E_{A,relax}$ due to the relaxation processes was determined from the linear fit of the dependence $ln\tau(T^{-1})$, according to equation (2.20). The values obtained corresponding to the temperature domains mentioned above are $E_{A,relax1} = 0.015\,eV$ and $E_{A,relax2} = 0.363\,eV$, which accords with the values obtained in [82] for samples of the same type ($MnFe_2O_4$).

The electrical conduction processes due to the transport of the charge carriers in the $MnFe_2O_4$ sample were analyzed based on the measurements of electrical conductivity. The appearance and behavior of AC conductivity can be explained using Jonscher's relation [80]:

$$\sigma = \sigma_{DC}(T) + \sigma_{AC}(\omega, T) \tag{2.21}$$

where σ_{DC} is the static component (DC conductivity) of electrical conductivity, being independent of frequency, but dependent on temperature; and σ_{AC} is the conductivity in alternating current (AC conductivity), which depends on both frequency and temperature. This component is correlated with the dielectric relaxation processes determined by the localized electric charge carriers and having the following expression [27, 83]:

$$\sigma_{AC}(\omega,T) = A(T)\omega^{n(T)} \tag{2.22}$$

where $A(T)$ is a temperature-dependent parameter and the exponent n is dimensionless and also temperature-dependent ($n<1$) [84].

Based on the experimental values of the Z′ and Z″ components of the complex impedance in fig. 2-18 and knowing the geometry of the sample, using relation (2.13) we can calculate the total conductivity σ, which is the same as the real component σ'. Figure 2-19 shows the frequency dependence of the total conductivity σ at different temperatures T for the investigated sample.

Figure 2-19 shows that for each temperature T, the conductivity σ remains constant at low frequencies until near 10 kHz; then, it increases sharply with increasing frequency. The conductivity spectrum in fig. 2-19, respects Jonscher's universal law (eq. 2.21).

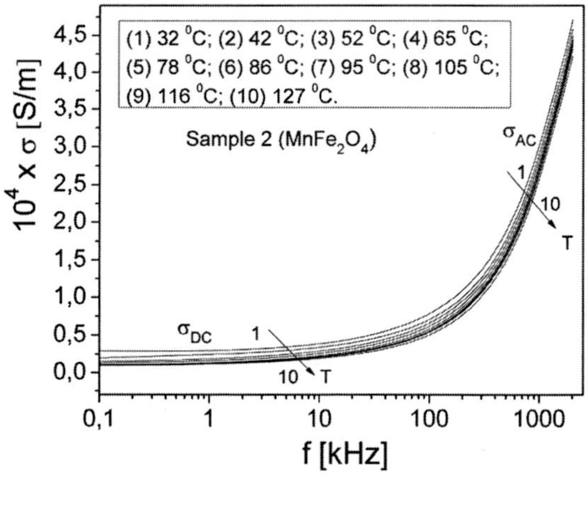

Fig. 2-19. Frequency dependence of conductivity σ at different temperatures for sample 2 ($MnFe_2O_4$).

This behavior can be explained using the Maxwell-Wagner model and the Koops phenomenological theory [85], which assumes that ferrites are systems of nanoparticles (grains) separated by thin layers of high strength (grain boundaries). As the frequency of the field increases, the ferrite nanoparticles become more active in terms of electrical conduction due to the jumping of electrons between the Fe^{2+} and Fe^{3+} ions [82, 86] so that the conductivity increases with increasing frequency.

Based on the values obtained for the static conductivity (fig. 2-19), at all temperatures T for which the measurements were performed, we assessed the temperature dependence of the static conductivity $\sigma_{DC}(T)$, which is shown in fig. 2.20 for the investigated sample.

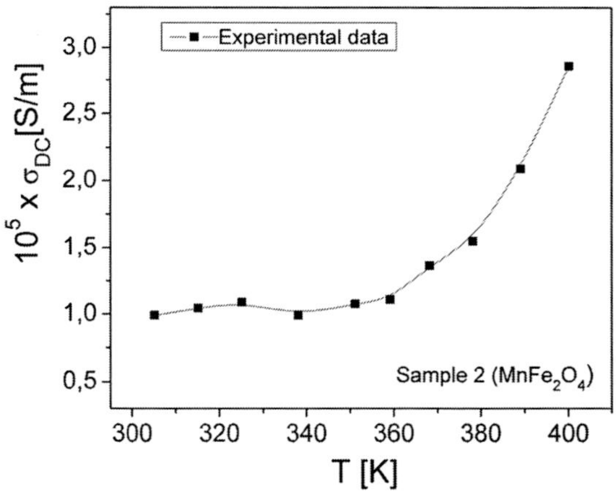

Fig. 2-20. Temperature dependence of the σ_{DC} conductivity of the $MnFe_2O_4$ sample.

Figure 2-20 shows that in the case of the investigated sample ($MnFe_2O_4$) the $\sigma_{DC}(T)$ component increases slightly as the temperature increases from 30 to about 65 °C, after which the increase is more pronounced at temperatures between 70 and 127 °C, indicating that the conduction process is thermally activated in the two temperature ranges. At the same time, the increase in σ_{DC} conductivity with temperature is due to the increase of drift mobility of the charge carriers, in accordance with the hopping-conduction mechanism [87]. Based on Mott's variable-range hopping model (VRH) [77], the temperature dependence of static conductivity σ_{DC} is described by the following equation:

$$\sigma_{DC} = \sigma_0 \exp\left(-\frac{B}{T^{1/4}}\right) \tag{2.23}$$

where σ_0 is the pre-exponential term and $B = 4E_{cond} / kT^{3/4}$, E_{cond} is the activation energy for electrical conduction [87, 88]. Based on equation (2.23), we can graphically represent the experimental dependence $\ln\sigma_{DC}$ (1/T $^{1/4}$) of the sample, shown in fig. 2-21.

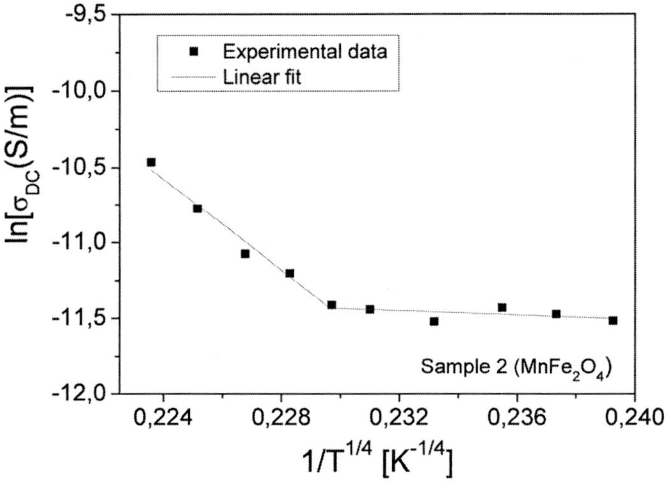

Fig. 2-21. Dependence $\ln\sigma_{DC}$ (1/T $^{1/4}$) for the $MnFe_2O_4$ sample.

Figure 2-21 shows a change in the slope of the curve for the investigated sample at temperatures of about 80 °C, due to both ferrite nanoparticles and particle separation limits [88, 89], as well as a possible structural changes in the sample at temperatures above 80 °C, determining the change in conductivity of the sample. By fitting the experimental dependence with a straight line $\ln\sigma_{DC}$ (1/T $^{1/4}$) in fig. 2-21, we obtained parameter B and could then calculate the activation energy for conduction.

Figure 2-22 shows the temperature dependence of the activation energy $E_{cond}(T)$ for the sample in the two temperature ranges.

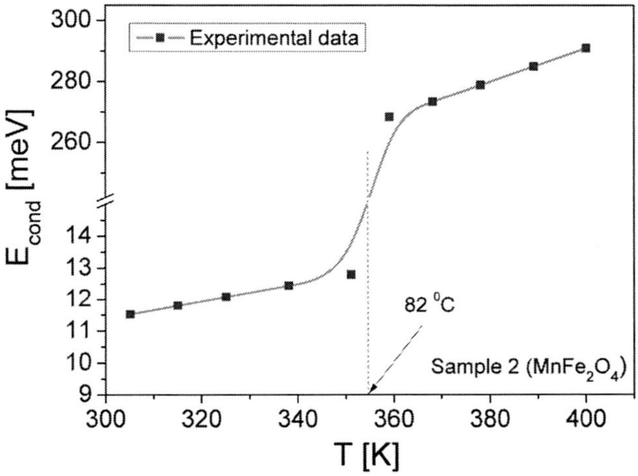

Fig. 2-22. Temperature dependence of the activation energy for conduction (E_{cond}) in the case of the $MnFe_2O_4$ sample in the temperature ranges 32-82 °C and 82-127 °C, respectively.

As can be seen in fig. 2-22, for both temperature ranges, E_{cond} increases linearly with increasing temperature, which indicates behavior similar to that of powdered materials [90] and is concordant with the VRH model, i.e. the electrical conduction in the sample investigated can be explained by the process of electrons hopping between localized states [91]. The values obtained for E_{cond} in the case of the investigated sample in the temperature range 32-82 °C are between 11 and 12.5 meV, and in the temperature range 80-127 °C they are between 0.260 and 0.295 eV (fig. 2-22), which is close to the value obtained for the activation energy due to the relaxation processes for similar temperature ranges. The results obtained for the activation energy due to conduction processes are similar to those obtained in [82, 88] for $MnFe_2O_4$ samples.

In the dispersion region corresponding to high frequencies ($f >$ *200* kHz, fig. 2-19), the conductivity σ_{AC} respects the equation (2.22), which, by logarithmic transformation, becomes:

$$\ln \sigma_{AC} = \ln A + n(T) \ln \omega \qquad (2.24)$$

Equation (2.24) shows a linear dependence between $ln\sigma_{AC}$ and $ln\omega$. Figure 2-23 shows the experimental dependence, $ln(\sigma_{AC})$ on $ln(\omega)$, for temperature in the high frequency range (f> 200 kHz), where equation (2.24) is valid.

By fitting the experimental dependence $ln(\sigma_{AC})(ln(\omega))$ in fig. 2-23 with a straight line, we can determine the exponent n corresponding to each temperature value T, based on equation (2.24).

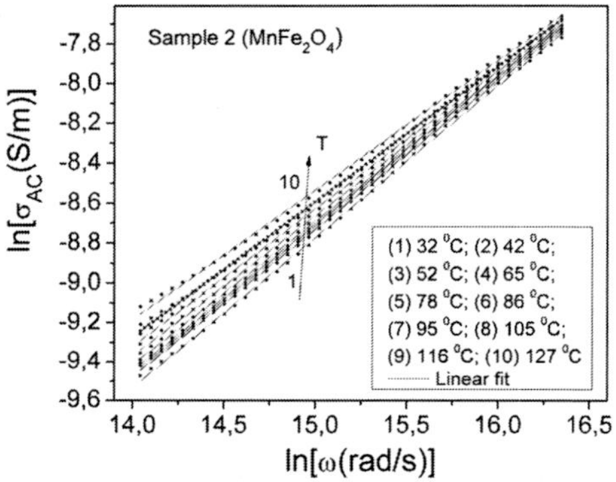

Fig. 2-23. Dependence of $ln(\sigma_{AC})$ on $ln(\omega)$ at different temperatures for sample 2 ($MnFe_2O_4$).

The study of the temperature variation of the exponent n provides us with information related to the electrical conduction mechanism in the investigated sample [92, 93]. In fig. 2-24, we present the temperature dependence of the exponent n for the ferrite sample $MnFe_2O_4$.

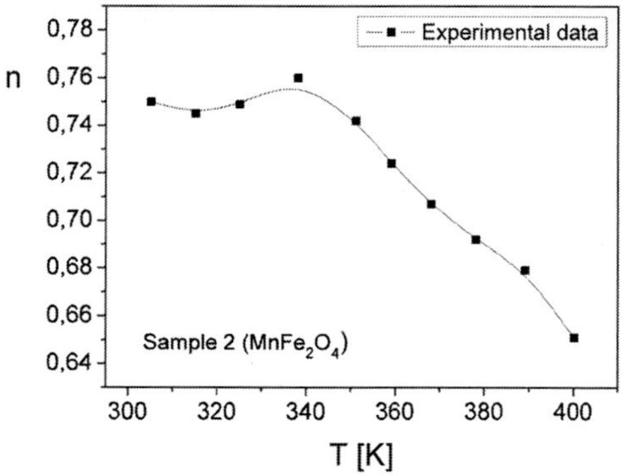

Fig. 2-24. Temperature dependence of exponent n for sample 2 ($MnFe_2O_4$).

In fig. 2-24, we can observe that the exponent n for the $MnFe_2O_4$ sample increases very little from 0.75 to 0.76 as the temperature increases from 30 to 70 °C and then decreases rapidly from 0.76 to 0.65 when the temperature rises in the range 70-128 °C. At the same time, we can observe in fig. 2-24 that the exponent n has relatively high values ($n > 0.5$), showing that the phenomenon of conduction is due to the hopping of charge carriers in the material between their nearest neighboring states, according to the CBH model [78]. According to this model, the exponent n can be approximately written [78] as:

$$n = 1 - \frac{6kT}{W_m} \tag{2.25}$$

where W_m represents the maximum energy of the barrier, which we consider equal to the energy of the band gap. Knowing the values n determined and presented in fig. 2.25 for the investigated sample, based on relation (2.25) we determined the energy of the barrier W_m at all the investigated temperatures. The results obtained are shown in fig. 2-25.

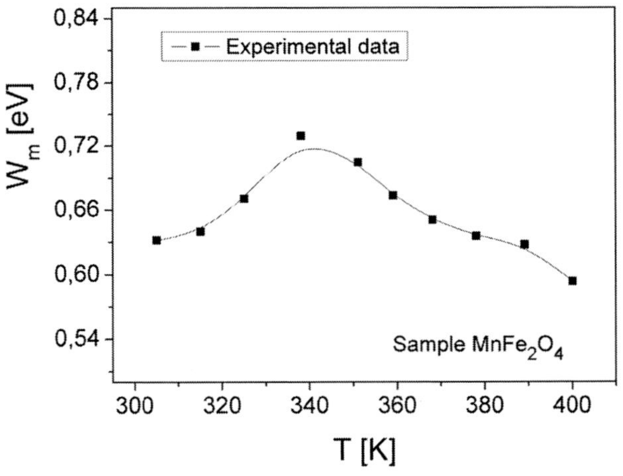

Fig. 2-25. Temperature dependence of the W_m barrier energy for sample 2 ($MnFe_2O_4$).

As can be seen in fig 2-25, in the case of manganese ferrite, the energy W_m has a maximum equal to 0.72 eV at a temperature of 70-80 °C. This correlates very well with the static conductivity σ_{DC}, which has a minimum at a similar temperature (see fig. 2-24). The value obtained for W_m (fig. 2-25) is in agreement with the values obtained by other authors [94, 95] for samples of manganese ferrite $MnFe_2O_4$ or manganese ferrite substituted by Cu^{2+} or Ca ions (with values between 1 and 2 eV).

2.8. AC Magnetic Measurements for Particulate Magnetic Oxides

Due to the small size of the embedded particles, granular magnetic materials such as metal oxide powders, composites, ferrofluids, or ferrite powders for use in permanent magnets, require fairly strong fields to reach magnetic saturation. Accordingly, instruments like a vibrating sample magnetometer (VSM) [96], alternating gradient magnetometer (AGM) [97], or superconducting quantum interference device (SQUID) magnetometer [98] are the most appropriate for making magnetic measurements of such materials due to their high field capability using electromagnets or superconducting coils. In addition, laboratory-made instruments like the Hall-effect magnetometer [99], the vibrating coil magnetometer [100], the Kerr-effect hysteresis loop tracer [101], or the flux extraction magnetometer [102], using electromagnets for field generation, as well as the open circuit AC induction hysteresigraph [103], using multilayer solenoid-type coils (field coils), are also of interest. Although this latter type of instrument is probably the most popular (due to its lower cost and sufficiently high precision), the field strength capability of such field coils rarely reaches the kOe range [104], thus making AC induction hysteresigraphs generally inappropriate for the characterization of materials with high coercivity (e.g. for permanent magnetic materials). Moreover, if only a small amount of magnetic material (e.g. less than 100 mg of final product, as commonly delivered by synthesis in the laboratory) is available for measurement, additional issues regarding air flux compensation, signal-to-noise ratio, and electronic circuit errors inevitably occur in strong fields. Solutions to such drawbacks have been proposed by the author in Ref. [105] for an AC induction hysteresigraph using a field coil with accurate sine waveform field and amplitude reaching 7 kOe (557 kA/m) at a frequency of 50 Hz, performing measurements on small amounts (down to 10^{-4} g) of particulate magnetic materials like ferrite magnet powders, nanopowders, ferrofluids, and nanocomposites.

2.8.1. The method

This hysteresigraph is built from three analog blocks: a passive electrical circuit containing the driving field coil and the primary signal detectors using electromagnetic induction, e.g. $v_h(t)$ and $v_m(t)$, and two electronic measuring channels, e.g. *channel H* and *channel M*, which provide two output signals, e.g. $u_H(t)$ and $u_M(t)$, as waveforms reproducing evolution over time $H(t)$ and $M(t)$ of the field and magnetization, respectively. These output signals are subjected to computer-assisted analog-to-digital conversion and data acquisition for further numerical processing and calibration.

The first analog block consists of a *field coil* circuit tuned to serial resonance by means of a power capacitor and two induction signal generators: a field sensing coil (c_h) thrat delivers the *field signal* $v_h(t)$ and a *compensated pick-up coil system* by means of which the *magnetization signal* $v_m(t)$ is obtained. This latter component is made from a search coil (c_m) containing the sample and compensation coil (c_c). The axial field generated by the *field coil* when the current passes along its turns $i_h(t)$ is:

$$H(x,t)=k_H(x)\cdot i_h(t) \tag{2.26}$$

where x is the coordinate along the axis of the *field coil* (let $x = 0$ in the centre) and $k_H(x)$ is a factor (sometimes called the *field constant*) that depends on the number of turns and geometry of the coil. However, with a large length-to-diameter ratio, the field is uniform in the central region with good approximation. Accordingly, if n_h, n_m, and n_c are the numbers of turns of the sensing coils c_h, c_m, and c_c, respectively, x_h, x_m, and x_c are their positions and S_{eh}, S_{em}, and S_{ec} are their *effective turn areas*. As such, if s is the cross-sectional area of the sample, the magnetic fluxes sensed by the coils will be:

$$\begin{aligned}
\Phi_h(t)&=\mu_0 n_h S_{eh} H(x_h,t)\\
\Phi_m(t)&=\mu_0 n_m\left[S_{em}H(x_m,t)+sM(x_m,t)\right]\\
\Phi_c(t)&=\mu_0 n_c S_{ec} H(x_c,t)
\end{aligned} \tag{2.27}$$

respectively, (μ_0 is the vacuum permeability) and the corresponding electromotive forces induced in these coils will be:

$$e_h(t) = -\mu_0 n_h S_{eh} \frac{dH(x_h,t)}{dt}$$

$$e_m(t) = -\mu_0 n_m \left[S_{em} \frac{dH(x_m,t)}{dt} - s \frac{dM(x_m,t)}{dt} \right]$$

$$e_c(t) = -\mu_0 n_c S_{ec} \frac{dH(x_c,t)}{dt} \tag{2.28}$$

It can be observed that the emf induced in the coil c_m is of the form:

$$e_m(t) = e_{mH}(t) + e_{mM}(t) \tag{2.29}$$

where the second term carries the information on the magnetization rate of change inside the sample and therefore should be extracted from the total signal $e_m(t)$. This is possible in principle, provided the following condition (the *compensation condition*) exists:

$$k_H(x_c)\, n_c S_{ec} = k_H(x_m)\, n_m S_{em} \tag{2.30}$$

In this case, the signals rendered by the field sensing coil c_h and the compensated pick-up system c_m-c_c will be:

$$v_h(t) = -\mu_0 n_h S_{eh} \frac{dH(x_h,t)}{dt}$$

$$v_m(t) = -\mu_0 n_m s \frac{dM(x_m,t)}{dt} \tag{2.31}$$

The signals $v_h(t)$ and $v_m(t)$, recorded for an Ni sample, are shown in fig.2-26.

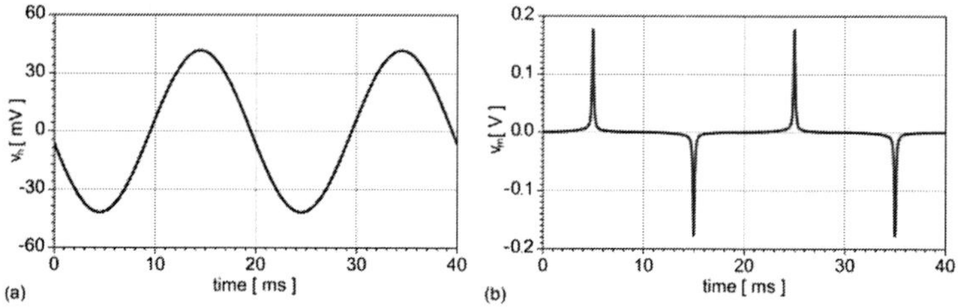

Fig. 2-26. Signals profiles of (2.31); the sample was a 0.090 mm diameter Ni wire.

Signals $v_h(t)$ and $v_m(t)$ will run through the stages of *channels H* and *M*, which exhibit similar architectures, i.e. they contain input protection (clipping) circuits, linear amplifiers ($u_h(t)$ and $u_m(t)$ as the output signals of these amplifiers, and analog integrators rendering the desired ouput signals $u_H(t)$ and $u_M(t)$. Figure 2-27 presents the waveforms of these output signals corresponding to the input signals in fig. 2-30.

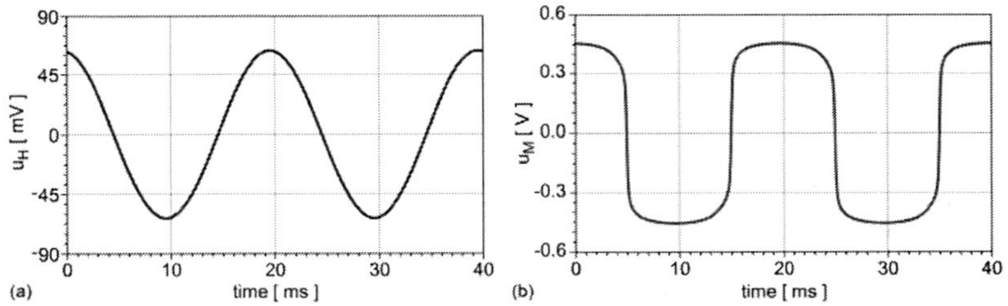

Fig. 2-27. Output signal profiles of (2.32); the sample was a 0.090 mm diameter Ni wire.

As will be described later, accurate waveform $u_M(t)$ signals require an additional correction circuit in which signals $u_h(t)$, $u_m(t)$, and $u_H(t)$ are combined. The main cause of this lies in the fact that expressions (2.31) for signals $u_h(t)$ and $u_m(t)$ were derived by assuming c_h, c_m, and c_c as ideal coils (i.e. having vanishing resistance, reactance, and stray capacitance). Under these simplified conditions, if g_H and g_M are the gains of the amplifiers and τ_H and τ_M are the time constants of the analog integrators of *channels H* and *M*, respectively, the output signals of these channels will be (note that the operational amplifier (op amp) based analog integrators cause sign inversion):

$$u_H(t)=\frac{1}{\alpha_H}H(t)$$

$$u_M(t)=\frac{1}{\alpha_M}M(t) \tag{2.32}$$

where:

$$\alpha_H = \frac{\tau_H}{\mu_o S_{eh} n_h g_H} \cdot \frac{k_H(0)}{k_H(x_h)}$$

$$\alpha_M = \frac{\tau_M}{\mu_o s n_m g_M (1 - q_{str})}$$

(2.33)

are the calibration factors of the two measuring channels. The *stray flux* factor $0<q_{str}<1$ occurs from a correction, taking into account the fact that the returning lines of the stray field generated by the magnetized sample itself intersect, in part, the turns of the search coil c_m, thus causing a decrease of the flux sensed by this.

By combining the two output signals $u_H(t)$ and $u_M(t)$, the profile of the hysteresis loop results (see fig. 2-28) as the geometrical locus of the synchronous points (u_H, u_M).

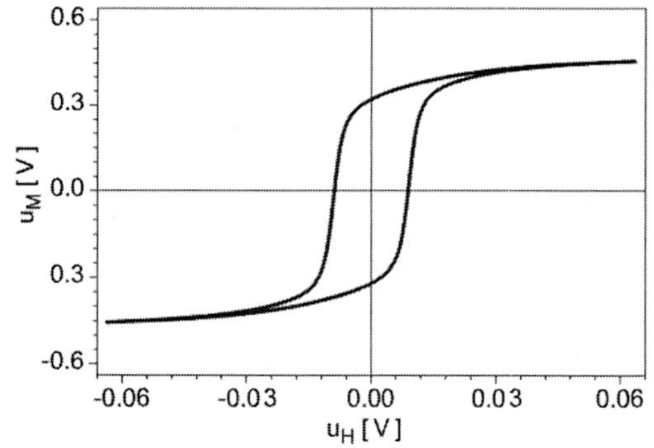

Fig. 2-27. Profile of the hysteresis loop built from signals $u_H(t)$ and $u_M(t)$ in fig. 2-26.

As far as the unwanted stray flux, e.g. $\Phi_{str}(t)$, originates from the magnetized sample. It may be expressed in terms of the intrinsic flux $\Phi_{in}(t) = \Phi_o sM(t)$ as:

$$\Phi_{str}(t) = -q_{str} n_m \Phi_{in}(t)$$

(2.34)

q_{str} may be evaluated through a magnetostatic calculus. In *Annex.2.8.1*, the case of a circular cross-section sample and pick-up coil is treated. As an example, one finds from this calculus (numerical) that, for a coil c_m of 25 layers and 200 turns per layer made from a tightly wound 0.07 mm thick wire on a 4 mm diameter spool, containing a 3 mm thick sample, q_{str} = 0.0775 for a 2 cm long sample; 0.0284 for a 3 cm long sample; and 0.0148 for a 4 cm long sample. Non-negligible errors in the determination of magnetization correspond to 8.4 %, 2.9 %, and 1.5 %, respectively.

Once the calibration factors α_H and α_M and the demagnetizing factor N_D (fluxmetric) of the sample [110] are known, the parametric equations of the real hysteresis loop from eq. (2.33) will be:

$$H(t)=\alpha_H u_H(t)-\frac{\alpha_M u_M(t)}{N_D^{-1}-1}$$

$$M(t)=\frac{\alpha_M u_M(t)}{1-N_D}$$

(2.35)

Finally, we note that *channel M* may be calibrated to the *specific magnetization* $\sigma = M/\rho$, where ρ is the mass density, in which case, the factor α_M will be replaced with:

$$\alpha_\sigma=\frac{\tau_M}{\mu_0 n_m g_M\left(1-q_{str}\right)\rho_l}$$

(2.35)

where ρ_l is the mass per unit length. It is worth noting that when non-homogeneous materials like powders, mixed powders, or composites etc. are measured, the quantity σ is preferred to M and should be regarded as an *apparent specific magnetization.*

Partly justified, the above approach is acceptable if materials that require moderate or weak saturation fields are examined [111-113]. However, if, as in our case, measurements are performed on small amounts of magnetic material under strong field conditions, the situation becomes more complicated. This is because air flux compensation now becomes more difficult and also $v_M(t)$ signals of acceptably large amplitude require a fairly large $sn_m g_M$ product. Several limitations to increasing this product occur (see Ref. [105] for details). Thus, on the one hand, to avoid the effect of the stray flux generated by the sample, its length should exceed that of the pick-up coil, which will limit the cross-sectional area s. Moreover, when saturation requires strong fields, samples of sufficiently large geometrical ratio are needed to limit the effect of the demagnetizing field, which also limits this area. On the other hand, the number n_m of pick-up coil turns will also be limited. A large value for this number will make the stray capacitance (C_{sm}) of the coil non-negligible, especially if thin winding wires are used to minimize the increase of the effective turn area S_{em}. Under these circumstances, the pick-up coil cannot be approximated more to an ideal emf generator, but rather to an *RLC* dipole having such a generator embedded. Fig. 2-28 gives the equivalent circuit of the coil [114].

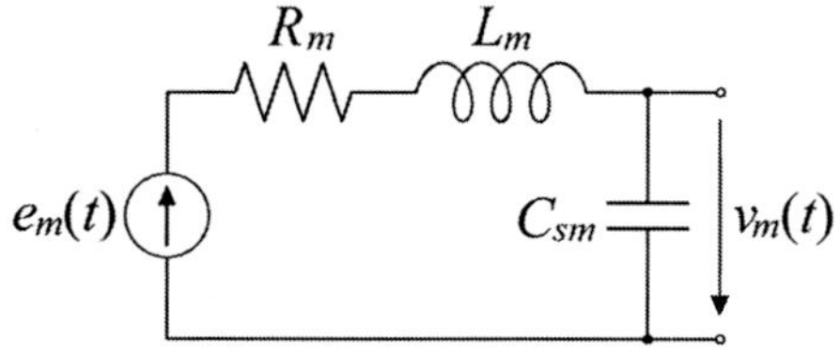

Fig. 2-28. The equivalent circuit of a real pick-up coil of resistance R_2, inductance L_2, and stray capacitance C_{sm}; for simplicity's sake, the case of pure $e_m(t)$ induction was considered.

If a signal of the form given in eq. (2.29) is induced in the search coil c_m, the output signal may still be written in the same form, but, to an extent depending on its spectral composition and on the electric parameters of the coil, the two components will be expected to exhibit waveform alterations (see Ref. [105] for details). Whereas only a slight amplitude decrease and phase lag could affect the contribution of $e_{mH}(t)$ to the output signal of c_m under sinusoidal field conditions, the profile of the contribution of $e_{mM}(t)$ may notably differ from that of $e_m(t)$, especially if the sample exhibits a steep-sided hysteresis loop, in which case the latter signal consists of sharp AC peaks and is rich in high-order odd harmonics. Moreover, as C_{sm} increases, the corresponding output peaks are splayed and delayed, the coil resonance frequency decreases, and parasitic self-oscillations may be triggered by the induced signal itself, any attempt to remove these by analog filtration inevitably leads to additional pulse shape alteration. The electrical parameters of the air flux compensation coil c_c will inherently differ from those of c_m, in which case a slight phase lag (frequency-dependent) is expected between the output signal and its field-associated component rendered by c_m. This will inevitably result in a mismatch of these two signals (even if the position of c_c had been carefully set) and the output signal of the compensated coil system should rather be regarded as:

$$v_{mr}(t) = v_m(t) + \delta e_h(t) \quad . \tag{2.37}$$

where $\delta e_h(t)$ is a parasitic signal that progressively affects the shape of the hysteresis loop as the amount of sample material decreases and the amplitude of the driving field increases. This issue may be solved for the sine waveform $H(t)$, since $\delta e_h(t)$ is also a sine waveform in this case. Cancellation of this parasitic signal may be done by subtracting a correction signal from the real output $u_{mr}(t)$, resulting from amplifying $v_{mr}(t)$, of the form:

$$u_{corr}(t)=\xi' u_h(t)+\xi'' u_H(t) \tag{2.38}$$

reproducing both in amplitude and in phase the signal $\delta u_h(t) = g_M \delta e_h(t)$. Since both signals, $u_h(t)$ and $u_H(t)$, are physically available, $u_{corr}(t)$ may be built from these signals by precisely setting the *tuning fractions* ξ' and ξ''. To this end, a solution based on a differential application of the operational amplifier (op amp) is proposed in fig. 2-29.

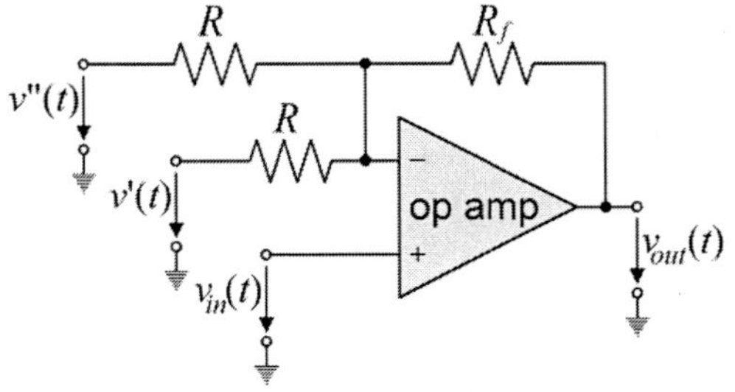

Fig. 2-29. Op amp-based differential amplifier with the inverting input as the summing point.

Expressing the feedback resistor as $R_f = \beta R$, if the following signals:

$$\begin{aligned} v_{in}(t)&=u_{mr}(t) \\ v'(t)&=\xi'\left(2+\beta^{-1}\right)u_h(t) \\ v''(t)&=\xi''\left(2+\beta^{-1}\right)u_H(t) \end{aligned} \tag{2.39}$$

are applied to the inputs of the circuit (the inverted input is configured as a summing point), the output signal will be:

$$v_{out}(t) = (2\beta + 1) u_m(t) \tag{2.40}$$

If, as with c_c, the *field sensing* coil c_h is also a flat coil placed at one end of *the field coil*, its number of turns n_c will be smaller than n_m, even by two orders of magnitude for $u_H(t)$ amplitudes in the 0.1-1 V range. Accordingly, c_h will be expected to exhibit very low stray capacitance and thus cause a negligible phase shift in the signal $u_H(t)$ with respect to $H(t)$ at a frequency of 50 Hz.

For g_M, if a low noise and low offset voltage op amp is used to build the linear amplifier of *channel M*, the gain should be limited to relatively low values (e.g. 5 and 9, which is selectable in our device) in order to prevent signal distortion. This is because both the gain and the phase shift of such circuits are frequency-dependent having unwanted influence on the waveform of the output signal. Clearly, the higher the gain is, the worse the frequency response. Another issue results from the fact that, in order to avoid output drift, op amp-based analog integrators like those used in the final stages of the measuring channels *H* and *M* require DC stabilization. To this end, a large resistor R_f is connected across the negative feedback capacitor C_f (see fig. 2-30). Since this will cause a discharge process, both the phase and the waveform of the output signal will be affected.

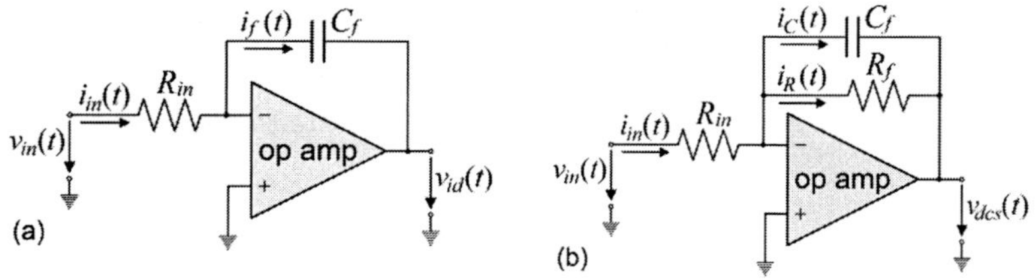

Fig. 2-30. The op amp-based analog integrator: a) ideal and b) dc stabilized.

If the output signals:

$$v_{id}(t) = -\frac{1}{\tau}\int v_{in}(t)\mathrm{d}t$$

$$\tau = R_{in}C_f\,;\tau_f = R_fC_f$$

$$v_{dcs}(t) = -\frac{1}{\tau}\int v_{in}(t)\mathrm{d}t + \frac{R_f}{\tau_f}\int i_R(t)\mathrm{d}t \tag{2.43}$$

from the ideal (fig. 2-30 a)) and the DC-stabilized (fig. 2-30 b)) integrator are compared, one can see that the cumulative effect of the parasitic current $i_R(t)$ constitutes the difference. Clearly, the weaker this current is, the smaller the integration error will be. For the *integrator H*, using the sine waveform $H(t)$, the DC stabilization will only cause a forward phase shift of $u_H(t)$, leading to an apparent increase in coercivity, which may be mistaken for a dynamic hysteresis effect caused by the field generated in the sample by eddy currents induced by reversals of magnetization. The situation is more complicated in the case of *integrator M* (an example is given in fig. 2-31) because the C_f discharge facilitated by the feedback resistor will cause a phase shift leading to the occurrence of secondary loops at the tips of the hysteresis loop, as well as to an apparent decrease in remanence. However, in fig. 2-31 we can see that a reasonable choice for τ_f may exceed the field period by three orders of magnitude. To what extent the above issues affect the accuracy of the measurement depends on the electrical characteristics of the pick-up coil c_m, of the *amplifier M*, and of the two *analog integrators*, as well as the sample geometry.

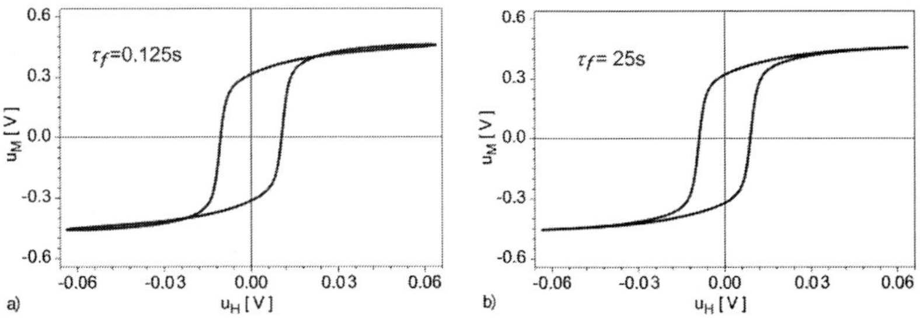

Fig. 2-31. The effect of DC stabilization of the analog integrator (R_{in} = 2.5 $k\Omega$, C_f = 2.5 μF) on the shape of the 50 Hz frequency hysteresis loop: a) R_f = 50 $k\Omega$ and b) R_f = 10 $M\Omega$. The sample is the same thin Ni wire as used in fig. 2-25.

2.8.2. The instrument

A description of the instrument, starting with its main components, is presented here. The magnetic field is generated by a multilayer *field coil* (solenoid-type) tuned to series resonance and supplied with sine wave AC currents from the 220 V/50 Hz line by means of a cascade built from a power autotransformer and a 220 V/2 kV step-up transformer. Alternatively, the frequency of the currents may be changed within reasonable limits, for example to between 20 and 100 Hz, by replacing the autotransformer connected to the 50 Hz power line with a commercial 2 kW audio amplifier driven by a sine wave oscillator and having the primary winding of a power step-up transformer rising to 220 V output voltage as the load. A field-sensing coil and a compensated pick-up coil system provide the voltages $v_h(t)$ and $v_m(t)$ as the input signals of the measuring *channels H* and *M*, respectively. *Channel H* has three stages: an input stage, made from an overvoltage protection circuit and a voltage follower; a second stage with a non-inverting linear amplifier; and an output stage with an analog integrator. *Channel M* exhibits similar architecture, except for the fact that a compensation circuit for the parasitic signal $\delta e_h(t)$ is included in the second stage. To this end, the solution in fig. 2-29 was adopted by applying the signal $v_m(t)$ at the non-inverting input and the components $v'(t)$ and $v''(t)$ of the correction signal (2.38) at the inverting input, now configured as a summing point (the latter signals may be applied either directly, or via individual phase-reversal by means of two inverting unity-gain amplifiers). Although expected to coincide with those in eq. (2.32), the output signals $u_{\mathrm{H}}(t)$ and $u_{\mathrm{M}}(t)$ may still be affected by the inherent noise and for $u_{\mathrm{M}}(t)$ by the limited precision in $\delta e_h(t)$ cancellation (manually setting fractions ξ' and ξ''), especially in the case of weak magnetic samples. Corrections to these errors are applied through numerical processing.

2.8.2.1. The field coil

The field coil was manufactured by winding approximately 32,500 turns of 0.3 mm enameled Cu wire, wrapped tightly on a 21 cm long and 1.2 cm outer diameter ceramic spool (fig. 2-32 a)). The field constant of the coil was determined with a 50 Hz sine wave driving current conditions, by measuring the emf induced in a 50 turn, 8.453 mm mean diameter single layer probe coil placed in the center of the *field coil*: $k_H = 157.2\ kAm^{-1}/A$ (1.976 kOe/A). By moving the coil along the axis of the *field coil* bore, the axial distribution of the field strength (fig. 2-32 a)) was also determined. In addition, from measurements on the frequency dependence of the impedance modulus Z_{fc} (fig. 2-32 b)), the electric parameters $L_{fc} = 8.98\ H$, $C_{fc} = 0.2\ nF$, and $R_{fc} = 541\ \Omega$ were calculated on the basis of the equivalent circuit in fig. 2-28. Using these data, negligibly small phase lag (less than $10^{-5} rad$) was found at 50 Hz between the driving current and the generated field (details on the *field coil* characterization are given in *Annex 2.8.2*). Clearly, serial resonance is needed to compensate the coil inductance for maximum field strength. However, since high AC voltage (exceeding 2 kV in amplitude) is required for maximum field strength, the field sensing coil solution was preferred to a non-inductive serial resistor for safety reasons.

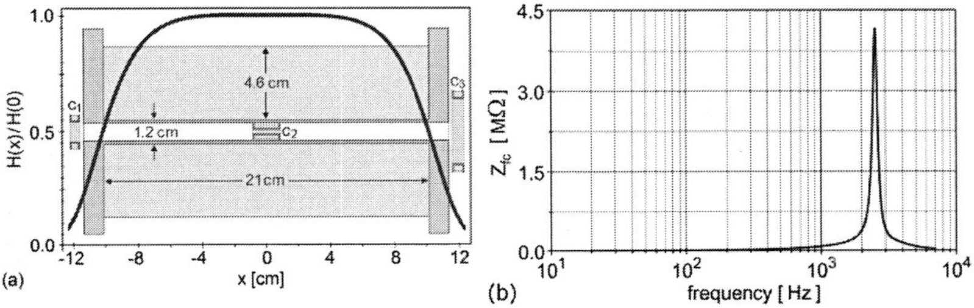

Fig. 2-32. The field coil: (a) the field strength axial dependence (in the background there is a sketch of the coil, including the three sensing coils); (b) the frequency dependence of the impedance modulus Z_{fc}.

2.8.2.2. The sensing coils

All the sensing coils are multilayer solenoids manufactured by uniformly winding a 0.07 mm diameter enameled Cu wire as shown in fig. 2-32 a). The pick-up coil c_2 is long and cylinder-shaped, while both the compensation and the field sensing coils c_3 and c_1 are flat and cylinder-shaped. Figure 2-33 shows the frequency dependence (see *Annex 2.8.3* for details on measurements) of the impedance modulus (fig. 2-33 a)) and of the equivalent serial inductance of these coils (fig. 2-33 b)). By means of a fitting procedure based on the equivalent circuit in fig. 4, the electrical parameters (R, L, C_S) and the resonance frequency (f_o) can all be determined from the recorded data. Table 2.8.1 gives the manufacturing details and electrical parameters of the coils (here, n is the number of turns; d_{in} and d_{out} are the inner and outer diameters; l is the length; and d_w is the diameter of the winding wire).

Table 2-4. Construction data and electrical parameters of the sensing coils.

coil	n	d_{in} [*mm*]	d_{out} [*mm*]	l [*mm*]	d_w [*mm*]	R [Ω]	L [*mH*]	C_S [*pF*]	f_o [*kHz*]
c_1	1200	16	24	4	0.1	158	28	17	227
c_2	5000	4	8	15	0.062	544	48	38	119
c_3	675	32	54	6.5	0.3	37.2	25	33	174

As expected, since the largest number of turns leads to the largest values for inductance and stray capacitance, the pick-up coil c_m exhibits the lowest resonance frequency.

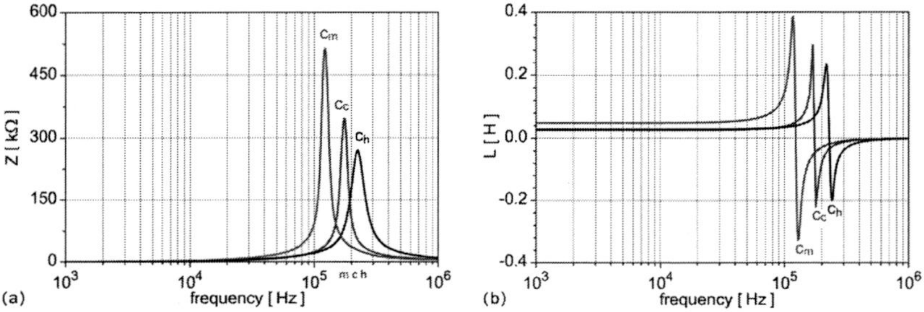

Fig. 2-33. Frequency dependence of the sensing coil parameters: a) the electrical impedance modulus Z and b) the equivalent serial inductance L.

2.8.2.3. The electric diagram

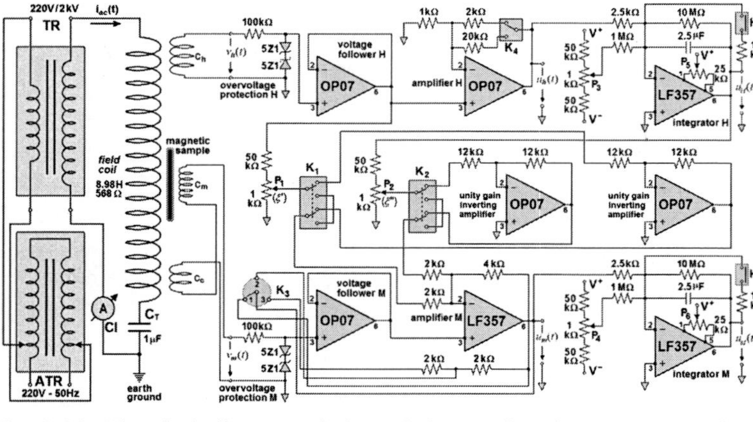

Fig. 2-34. Electrical diagram of the AC hysteresigraph: *ATR*: 220 V/50 Hz autotransformer; *TR*: 220 V/2.5 kV step-up transformer; *FC*: field coil; i_{ac}: AC driving current; C_r: tuning capacitor; *CI*: control instrument; 5Z1: 5.1 V voltage breakdown Zener diodes; K_1, K_2: $v'(t)$, $v''(t)$ phase inverting keys: P_2, P_3: ξ', ξ'' fraction tuning potentiometers; K_1, K_4: g_H, g_M selection keys; K_5, K_6: integrators reset keys; P_1, P_4: offset nulling potentiometers; OP-07 (Precision Monolithics): ultralow offset op amp; LF-357 (National): low offset JFET-input op amp; and V^+,V^-: DC supply voltages. All the resistors are carbon film-type selected with precision of 0.2 %, while all the potentiometers are 10 turns and helipot-type.

The electrical diagram of the hysteresigraph is shown in fig. 2-34 with two distinct blocks: the driving-field block and the signal-measuring block (composed of *channels H* and *M*), the former being completely isolated from the latter.

The field coil is supplied with 50 Hz AC currents $i_{ac}(t)$ from a cascade built using a 220 V/2.5 kV step-up transformer (*TR*) and a power autotransformer (*ATR*) connected to the 220 V single phase line. The coil circuit is tuned to series resonance for maximum field capability by means of a power capacitor (C_r). The measurements, as illustrated in fig. 2-35 a) by the frequency dependence of the circuit admittance, have indicated 17 dB attenuation at 100 Hz; 21 dB attenuation at 150 Hz; and a quality factor Q of close to 5. Accordingly, as the cross-sections of the *ATR* and *TR* cores are sufficiently large to maintain these cores far below magnetic saturation, a waveform of $i_{ac}(t)$ and, hence, a value of $H(t)$ very close to a sine waveform, will be generated. This is proven by the example in fig. 2-35 b), where the waveform of the current required for the generation of a 6 kOe magnetic field (the *ATR* now being supplied from the 220 V/50 Hz line) is compared to the associated plots of several user-defined fitting functions built of several first odd harmonics:

$$i_f(t)=I_{f,0}\sin\left(\omega t+\varphi_0\right)+I_{f,1}\sin\left(3\omega t+\varphi_1\right)+\ldots+I_{f,l}\sin\left[(2l+1)\omega t+\varphi_l\right] \tag{2.44}$$

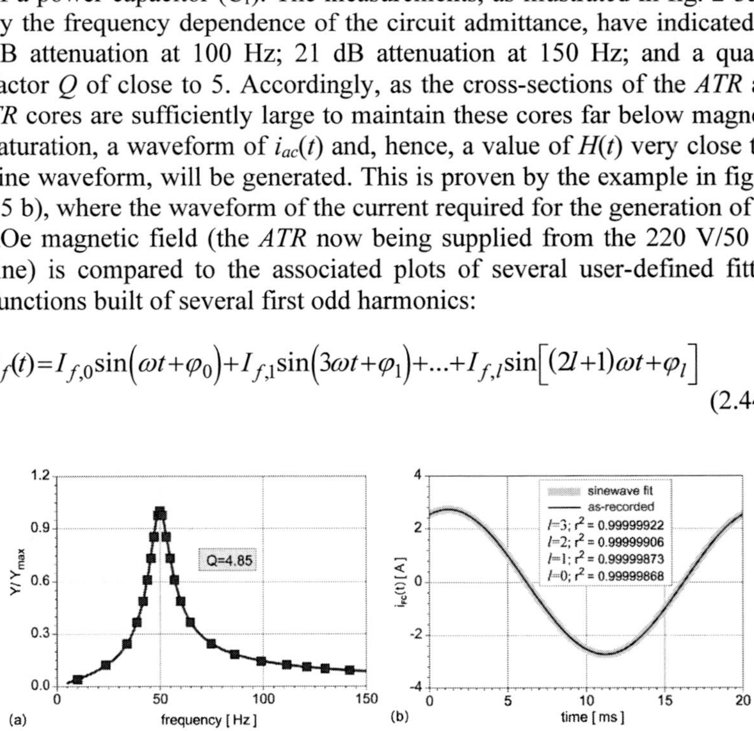

Fig. 2-35. *Field coil* circuit (tuned to 50 Hz resonance) characterization: a) the frequency dependence of the normalized admittance Y/Y_{max} (where Q is the quality factor) and b) a comparison of the waveform of $i_{ac}(t)$ required for the generation of a 6 kOe AC field (the AC current was recorded as the voltage drop across a 1.010 Ω non-inductive serial resistor) with the plots of the fitting function (2.44) for l = 0, 1, 2, and 3.

As can be seen, the plots for l = 0, 1, 2, and 3 coincide. Moreover, the high values of the coefficient r^2 of determination (with r^2 = 1 meaning a perfect match between the fit and the data points) confirm a profile that is very close to a pure sine wave, even for $I = 0$, of $i_{ac}(t)$ and, thereby, of $H(t)$. In other words, the 3ω, 5ω, 7ω, etc. components are negligibly small.

Each of the two overvoltage protection stages (*H* and *M*) consists of a clipping circuit built from a quadrupole made of a 100 kΩ current limiting resistor, two 5.1 V breakdown voltage Zener diodes connected in series opposition, and a voltage follower using an OP-07 (PMI) op amp of 1 MHz gain-bandwidth product (GBP). Note that as long as signals $v_h(t)$ and $v_m(t)$ do not exceed the breakdown voltage of the Zener diodes, the overvoltage protection circuits are not activated, which means that the input impedance of the two measuring channels remains close to 80 MΩ. In other words, both the field sensing coil and the pick-up coil system practically work in an open circuit, which ensures that the signals directed to the input of *amplifiers H* and *M* will be exactly $v_h(t)$ and $v_m(t)$, respectively. For *amplifier H* (with gain commutable between 3 and 21), no phase lag between the input and output signals was detected up to the 10 kHz range. This also held for the OP-07 (PMI) op amp-based inverting unity gain amplifiers used in building the signal for $\delta_{eh}(t)$ correction. For *amplifier M*, because under 50 Hz field conditions the frequency spectrum of $v_m(t)$ is expected to extend up to the 100 kHz range for samples of a steep-sided hysteresis loop, higher GBP op amps are required to prevent

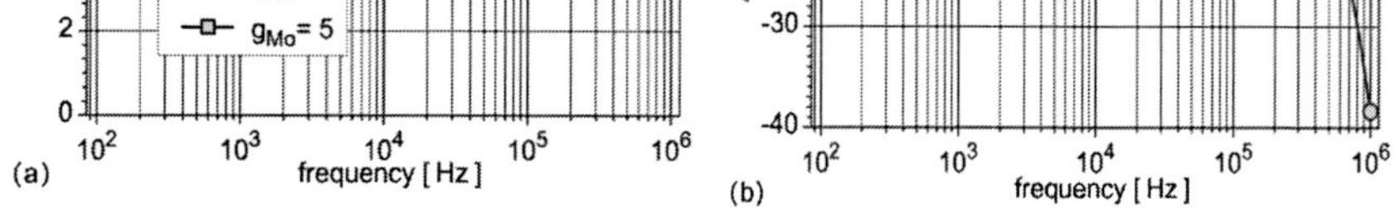

Fig. 2-36. The frequency response of the LF-357 op amp-based *amplifier M*: (a) the gain and (b) the output signal phase shift $\Delta\varphi_M$. 5.01 and 9.01 g_M values and -0.009 *rad* and -0.039 *rad* maximum phase delay resulted. In (b), the lines are plots of the fitting function (2.45) using the parameters listed in table 2-4.

(or at least to limit) gain decay and output signal phase shift (both effects, increasing with frequency, cause unwanted waveform alteration). To this end, the 20 MHz GBP LF-357 (NATIONAL) op amp was chosen, as sine waveform signal tests (details given in *Annex 2.8.2*) have shown (see fig.

2-41) that a gain below 20 dB (g_M <10) is practically constant up to 100 kHz, while the phase shift does not exceed 0.01 *rad* for $g_M = 5$ and 0.04 *rad* for $g_M = 9$.

It was found that logistic-type frequency dependence:

$$\vartheta(\omega) = -\alpha\left[1+\left(\frac{\beta}{\omega}\right)^{\eta}\right]^{-1} \tag{2.45}$$

of the gain-dependent parameters α, β, and η, as listed in table 2-5, fitted the phase shift data in fig. 2-36 b) fairly well.

Table 2-5. The gain-dependent parameters α, β, and η.

g_M	α [*rad*]	β [s^{-1}]	η
5	47.696	$1.2205 \cdot 10^8$	1.684
9	4.9471	$2.6460 \cdot 10^7$	1.283

As will be shown further on, to a large extent the choice of the LF-357 op amp meets the low distortion requirements for our sharp AC pulse input signals.

Integrators H and *M* (identical) were built using the same J-FET input op amp LF-357 (NATIONAL) and τ_f was set close to 25 *s*. As will also be shown later (*Section 2.8.2.4*), this is sufficiently large to avoid any detectable phase lag of $u_H(t)$, or waveform distortion of $u_M(t)$. As the gains of *amplifiers H* and *M* were set to low values, offset voltage nulling was only applied to the final stages. For computer-assisted analog-to-digital conversion and data acquisition of signals $u_H(t)$ and $u_M(t)$, a 16 *bit* resolution, 150 KS/*s* sampling rate parallel (simultaneously-triggered channels) DT-9816A (DATA TRANSLATION) card was used.

2.8.2.4. Signal processing and calibration

As strong driving fields were applied, no numerical correction was needed for the output signal of *channel H*, except for a fine offset voltage nulling (a residual offset voltage, in the mV range is inevitable from manual operation of the potentiometer P_5). This is because the amplitude of $u_H(t)$ is sufficiently large to mask the always present noise, which is in the mV range here. However, if necessary, the recorded signal $u_H(t)$ may accurately be denoised through sine wave-based fitting and the generation of a numerical signal.

For the output signal of *channel M*, a small offset voltage correction (similar to the case of *channel H*) is also the only operation required if the amplitude of $u_M(t)$ is in the volt range (when measuring normal magnetic materials like magnetite, maghemite, and ferrite powders, etc.). However, if this signal is weak (e.g. down to tens of mV), resulting from samples of low magnetic material content, an unwanted background signal of comparable level occurs increasingly affecting the profile of $u_M(t)$ as the amount of magnetic material content in the sample decreases. Accordingly, the output signal of *channel M* can be regarded as:

$$u_M(t)=u_{M,real}(t)+u_b(t) \tag{2.46}$$

where the background signal $u_b(t)$ includes amplified and integrated $\delta_{eh}(t)$ traces (remaining from manual correction); a low frequency (50 Hz and several odd and even harmonics) emf induced by stray fields generated by strong AC currents flowing through the line network; and also noise (in the mV range) from radio wave detection. In addition, a "burst" signal may be induced by current pulses generated by sparks at the brush-commutator contacts of the power electric motors operating at the time of measurement. Two representative $u_M(t)$ signal examples are given in fig. 2-37: a strong signal from a $Co_{0.4}Zn_{0.6}Fe_2O_4$ ferrite powder sample [115] (fig. 2.42 a)) and a weak signal from a carbon-based powder of low Fe_3O_4 content [116] (fig. 2-37 b)). The background signals $u_b(t)$, recorded after the removal of the sample from the pick-up coil, have also been plotted in both cases. While signals $u_M(t)$ and $u_{M,real}(t)$ practically coincide in the former case, subtraction of $u_b(t)$ from $u_M(t)$ will not ensure extraction of the desired $u_{M,real}(t)$ signal in the latter, which is expected since the noise is random and the two signals were not simultaneously, but rather consecutively recorded. The situation is depicted in fig. 2.43 b), in which the subtracted signal has also been plotted:

$$\Delta u_M(t)=u_M(t)-u_b(t)=u_{M,real}(t)+u_n(t) \tag{2.47}$$

where $u_n(t)$ is the noise signal. Clearly, the signal $\Delta u_M(t)$ should be denoised in such cases and to this end $u_{M,real}(t)$ reconstruction via numerical fitting based on the periodic functions of a tunable waveform will be preferable to low-pass filtering at the input of the DAQ card, or standard numerical denoising procedures like Fourier filtering, Savitzky-

Golay or spline smoothing, and eigen decomposition etc. This is because such signal conditioning procedures can be expected to alter the shape of the hysteresis loop, especially when steeply branched. As a more convenient alternative, fitting functions of the form below are considered.

$$u_{M,fit}(t)=\mathcal{U}_M\upsilon[\zeta(t)] \tag{2.48}$$

Here, Υ_M is a constant and $\upsilon[\xi(t)]$ is a composed function built from a transition-type symmetric function $\upsilon(\xi)$, whose variable ξ is a periodic symmetric function of amplitude Z playing the role of the transition time scaling factor, while the initial phase Ψ points to the zero intercept moment:

$$\zeta(t)=\mathcal{Z}\sin(\omega t+\gamma) \tag{2.49}$$

Among possible $\upsilon(\xi)$ forms, *Langevin*, *hyperbolic tangent*, *asymmetric sigmoid*, *Lorentz cumulative*, or *cumulative double sigmoid* functions may be considered (see *Annex 2.8.2.2*). However, as the amplitude of $\Delta u_{\mathrm{M}}(t)$ increases, the signal-to-noise ratio also increases and thus a better fit may be given by a superposition of functions (2.48). For example:

$$\upsilon_{M,fit}(t)=\sum_{i=1}^{n_M}\mathcal{U}_{M,i}\upsilon_i[\zeta_i(t)] \tag{2.50}$$

Here, n_M is a small integer (for relevant examples regarding this procedure, the reader is directed to Ref. [105]).

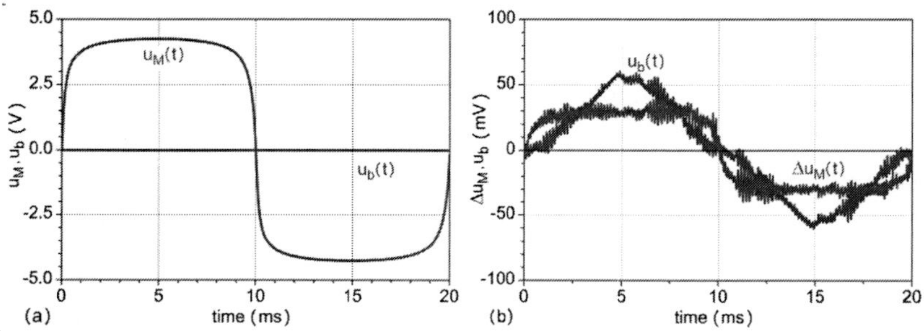

Fig. 2-37. Magnetization signals recorded for: a) a $Co_{0.4}Zn_{0.6}Fe_2O_4$ powder sample (strong signal, invisible noise) and b) a low Fe_3O_4 content powder sample (weak signal). The signal $\Delta u_M(t)$ is also shown.

The advantage of extending n_M becomes evident when materials containing different magnetic phases are examined, as illustrated in fig. 2-37 for an electrodeposited Ni tubular layer in which, according to Ref. [117], magnetic domains of transverse and longitudinal anisotropy leading to different magnetic behaviors coexist.

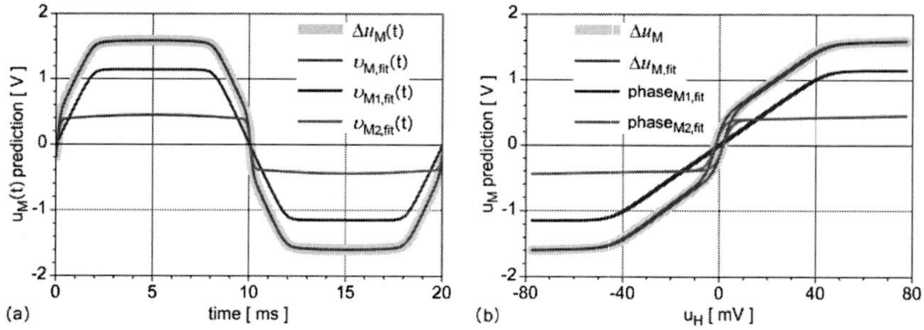

Fig. 2-38. a) Example of $\Delta u_M(t)$ fitting for an Ni layer containing two distinct magnetic phases; a) the fitting function $\upsilon_{M,fit}(t)=\upsilon_{M1,fit}(t)+\upsilon_{M2,fit}(t)$ is a *cumulative double sigmoid* function (see *Annex 2.8.1*); b) the associated hysteresis loops.

For *channel H* calibration, voltage signals $u_H(t)$ were recorded for each point during $k_H(x)$ determination, and using the signal amplitudes from sine wave fitting, the α_H values 66.8 kAm^{-1}/V (at 0.2 kAm^{-1}/V precision), or 840 Oe/V (at 3 Oe/V precision) were found for $g_H = 3$, and 9.55 kAm^{-1}/V (0.3 kAm^{-1}/V precision), or 120.0 Oe/V (at 0.4 Oe/V precision) for $g_H = 21$.

For *channel M*, we first note that factor α_M depends on the cross-sectional area of the sample, while factor α_σ depends on its mass per unit length. Rather, the following sample-independent factor may be considered:

$$\alpha_\Phi = \frac{\tau_M}{n_2 g_M} \tag{2.51}$$

for which factors α_M or α_σ (also including, depending on the case, corrections for the demagnetizing field and for the stray flux generated by the sample) result for individual samples from division by $\mu_o s$ and $\mu_o \rho_L$, respectively. In other words, provided n_2, τ_M, and g_M are known with sufficient precision (using high resolution instrumentation in electronic determination), *channel M* will be calibrated for the *intrinsic flux* (the contribution of the sample to the total flux). Accordingly, the following values: 0.251 μWb/V (2 nWb/V precision), or 25.1 Mx/V (0.2 Mx/V precision) at $g_M = 5$, and 0.139 μWb/V (1 nWb/V precision), or 13.9 Mx/V (0.1 Mx/V precision) at $g_M = 9$ resulted from our design data for α_Φ. Clearly, the final precision in the subsequent evaluation of α_M or α_σ for individual samples will also depend on the precision in measuring the cross-sectional area s, or the mass per unit length ρ_L. In fig. 2-39, two examples based on standard laboratory instrumentation (see *Annex 2.8.1*) are given: a $Mn_{0.4}Zn_{0.6}Fe_2O_4$ ferrite-based ferrofluid column [118] of 3.02 mm diameter and a 47.4 mm high Fe_3O_4 microcrystal column weighing 0.228 g [119].

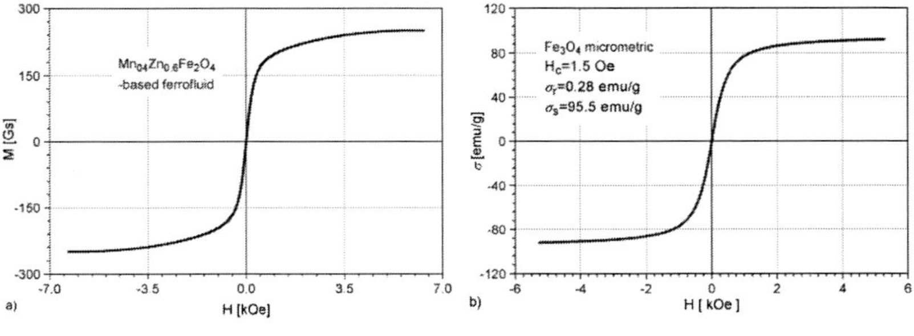

Fig. 2-39. Two examples of calibrated magnetic hysteresis loops: a) a $Mn_{0.4}Zn_{0.6}Fe_2O_4$ ferrite-based ferrofluid column 3.02 mm in diameter and a 47.4 mm high Fe_3O_4 microcrystal column weighing 0.228 g.

As can be seen, technical saturation is reached in both cases, which confirms the utility of our instrument for investigating particulate magnetic media.

2.8.3. Discussion

In order to evaluate the capability of the proposed instrument to render realistic hysteresis loops, in other words to quantify the possible frequency and gain-dependent errors, we seek to carry out a numerical simulation based on the design data. First, we note that if a 50 Hz pure sine waveform magnetic field is applied, no detectable phase shift or amplitude decay will be expected along the stages of *channel H*. Secondly, taking into account the spectral composition of the emf component of interest $e_m(t)$, the run of this signal through the stages of *channel M* needs to be examined via time-domain analysis. To this end, a virtual test signal, e.g. $e_{mv}(t)$, will be generated as a digital replica of $e_m(t)$, and the predicted signals, e.g. $v_{mv}(t)$, $u_{mv}(t)$ and $u_{Mv}(t)$, at the output of the pick-up coil, *amplifier M*, and *integrator M*, respectively (afterwards referred to as *realistic signals*), will be compared to their *ideal* versions:

$$v_{mv}(t)_{id} = e_{mv}(t)$$

$$u_{mv}(t)_{id} = g_M e_{mv}(t)$$

$$u_{Mv}(t)_{id} = -\frac{g_M}{\tau_M} \int e_{mv}(t) \mathrm{d}t$$

(2.52)

These are calculated by assuming an error-free (ideal) pick-up coil system and *channel M*, with the relative error signals below, regarded as a measure of the accuracy of our hysteresigraph design:

$$\delta v_{mv}(t) = \frac{[v_{mv}(t)_{id} - v_{mv}(t)]}{v_{mv}(t)_{id,max}}$$

$$\delta u_{mv}(t) = \frac{[u_{mv}(t)_{id} - u_{mv}(t)]}{u_{mv}(t)_{id,max}}$$

$$\delta u_{Mv}(t) = \frac{[u_{Mv}(t)_{id} - u_{Mv}(t)]}{u_{Mv}(t)_{id,max}}$$

(2.53)

As a form of $e_{mv}(t)$ suitable for time-domain analysis of AC circuits, a flexible combination of circular trigonometric functions to odd powers if the case shifted to reproduce asymmetric pulses, will be considered in the first step:

$$e_{mv}(t)=-\sum_{j=1}^{j_{max}} E_j \cos^{2N_j+1}\left[\omega(t-\delta t_j)\right]$$

(2.54)

where $\{N_j\}$ is an ordered set of custom-selected natural numbers ($N_1<N_2<...<N_{max}$); E_j is the amplitudes; and δt_j is the shift times.

In the second step, using the classical expansion of the cosine function to odd powers to a sum of cosines of odd multiple angles [120, 121], the spectral expansion of signal (2.54) will be brought to the form (see *Annex 2.8.3*):

$$e_{mv}(t)=-\sum_{j=1}^{N_{max}} E_{m,k} \cos\left[\omega_k(t+\gamma_k)\right] ,$$

(2.55)

where $k = 0, 1, \ldots N_{max}$ is the new summation index; $E_{m,k}$ is the amplitudes; and γ_k is the initial phases of the spectral components $\omega_k = (2k+1)\omega$, all calculable in terms of the given quantities ω, N_j, E_j, and δt_j. Then, $e_{mv}(t)$ will be induced in the pick-up coil c_m and the expected signal at its output, calculated on the basis of the equivalent circuit in fig. 2.33, will be:

$$v_{mv}(t)=-\sum_{k=0}^{N_{max}} V_{m,k}(\omega_k) \cos\left[\omega_k t+\gamma_k-\theta_k(\omega_k)\right] ,$$

(2.56)

where

$$V_{m,k}(\omega_k)=\frac{1}{\sqrt{\omega_k^2\tau_m^2+\left(1-\omega_k^2\omega_{mo}^{-2}\right)^2}}E_{m,k}$$

$$\text{with} \quad \tau_m=R_m C_{sm} \quad \omega_{mo}=\frac{1}{\sqrt{L_m C_{sm}}}$$

$$\theta_k(\omega_k)=\arctan\left(\frac{\omega_k\tau_m}{1-\omega_k^2\omega_{mo}^{-2}}\right)$$

(2.57)

Here, C_{sm}, L_m, and R_m are the stray capacitance, the inductance, and the winding resistance of coil c_m, respectively. Clearly, as C_{sm} increases, the initial phases $\varphi_k(\omega_k)$ gradually increase, while the amplitudes $U_{m,k}(\omega_k)$ gradually decrease, possibly distorting the $u_{mv}(t)$ waveform compared to $e_{mv}(t)$. For *amplifier M*, we found that g_M is practically constant up to 100 kHz, while the output signal phase shift is both gain and frequency-dependent. Then, denoting this phase shift for fixed g_M by $\Phi_k(\omega_k)$, the predicted output signal will be:

$$u_{mv}(t)=-\sum_{k=0}^{N_{max}} U_{m,k}(\omega_k)\cos\left[\omega_k t+\gamma_k-\varphi_k(\omega_k)\right] \tag{2.58}$$

with

$$U_{m,k}(\omega_k)=g_{\mathrm{M}}V_{m,k}(\omega_k)$$

$$\varphi_k(\omega_k)=\theta_k(\omega_k)+\phi_k(\omega_k) \tag{2.59}$$

Accordingly, if signal (2.58) is applied to the input of *integrator M*, it will render:

$$u_{Mv}(t)=\sum_{k=0}^{N_{max}} U_{M,k}(\omega_k)\sin\left[\omega_k t+\gamma_k+\psi_k(\omega_k)\right] \tag{2.60}$$

with

$$U_{M,k}(\omega_k)=\frac{1}{\tau_{\mathrm{M}}\sqrt{\omega_k^2+\tau_f^{-2}}}U_{m,k}(\omega_k)$$

$$\psi_k(\omega_k)=\arctan\left[\frac{1-\omega_k^2\omega_{mo}^{-2}-\omega_k^2\tau_f\tau_m}{\omega_k\left(\tau_f+\tau_m\right)-\omega_k^3\tau_f\tau_m}\right]+\phi_k(\omega_k) \tag{2.61}$$

Once $e_{mv}(t)$ is built, signals (2.56), (2.58), and (2.60) may be calculated using the electric parameters of the pick-up coil c_m, and *channel M* amplifier and integrator; a severe test signal, consisting of sharp AC pulses, like the signal rendered by the compensated pick-up coil system in fig. 2.30 b) will be considered (note that the induced emf component of interest $e_{mv}(t)$ is, in fact, inaccessible). Such a signal will thus be built using the numerical values listed in table 2-6.

Table 2-6. The parameters used in building the test signal $e_{mv}(t)$.

index j	N_j	E_j[V]	$\delta t_j[ms]$	index j	N_j	E_j[V]	$\delta t_j[ms]$
1	0	0.004	5.00	4	50	0.070	5.42
2	1	0.006	5.10	5	300	0.120	5.4
3	5	0.020	5.20	6	1000	0.780	5.44

In fig. 2-44, the ideal signals $v_{mv}(t)_{id}$ and $u_{Mv}(t)_{id}$ are compared to the realistic virtual signals $v_{mv}(t)$ and $u_{Mv}(t)$, respectively. Visually, we can observe the coincidence of these signals (this also holds for signals $u_{mv}(t)_{id}$ and $u_{mv}(t)$, not plotted here). In addition, fig. 2-45 confirms the visual coincidence between the ideal and realistic virtual hysteresis loops generated by means of a normalized sine waveform *field signal* $u_H(t)_{id}$, having the same phase advance as the real field signal in fig. 2.31 a). Clearly, some differences may still exist, but the calculus has shown that they are less than 0.035 % for $\delta v_{mv}(t)$; less than 0.025 % for $\delta u_{mv}(t)$; and less than 0.05 % for $\delta u_{Mv}(t)$. In other words negligibly small (see Ref. [105] for details).

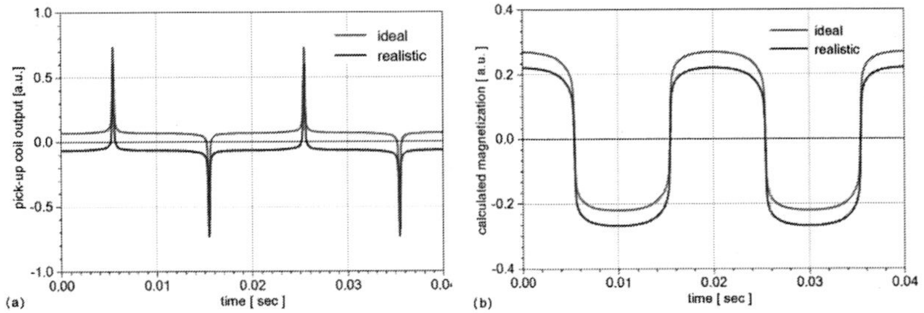

Fig. 2-40. Calculated 100 kHz bandwidth ideal and realistic signals (vertically shifted for visual ease) using table 2.8.3: a) at the pick-up coil output and b) at the output of *integrator M*.

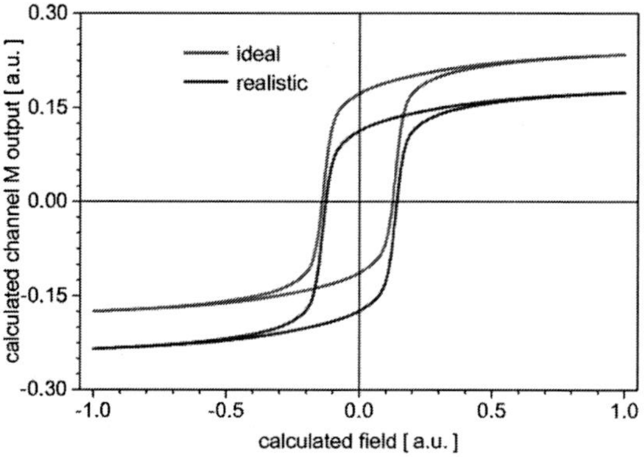

Fig. 2-41. Calculated 100 kHz bandwidth ideal and realistic hysteresis loops, and the relative error signals caused by the run of signal (2.55) across the stages of *channel M.*

These results prove that the proposed instrument will be able to render hysteresis loops, even steep sided ones, without detectable alteration in shape.

Annex 2.8.1.

Consider a circular cross-section bar (fig. 2-42 a)) of length $2h$ and radius a, surrounded by the turns of a coil of n_l layers and with n_t turns per layer. The radius of the turns in the first layer is b and the diameter of the winding wire is d; the bar is assumed to be uniformly magnetized, lengthwise. In order to evaluate the stray magnetic flux sensed by the coil, consider an elementary volume (fig. 2-42 b)), located inside the bar, around point P' of position $\boldsymbol{r}'(x',y',z')$ relative to the Cartesian frame $Oxyz$. If $\boldsymbol{M}_{bar}$ is the magnetization of the bar, the magnetic moment of this elemental dipole is:

$$\mathrm{d}\boldsymbol{\mu}' = \boldsymbol{M}_{bar}\,\mathrm{d}x'\,\mathrm{d}y'\,\mathrm{d}z' \tag{2.63}$$

and its contribution to the scalar magnetic potential in point P of position $\boldsymbol{r}(x,y,z)$ outside the bar will be [122]:

$$\mathrm{d}V_m(\boldsymbol{r},\boldsymbol{r}')=\frac{\mu_0}{4\pi}\frac{\boldsymbol{r}-\boldsymbol{r}'}{|\boldsymbol{r}-\boldsymbol{r}'|^3}\mathrm{d}\boldsymbol{\mu}'$$

(2.64)

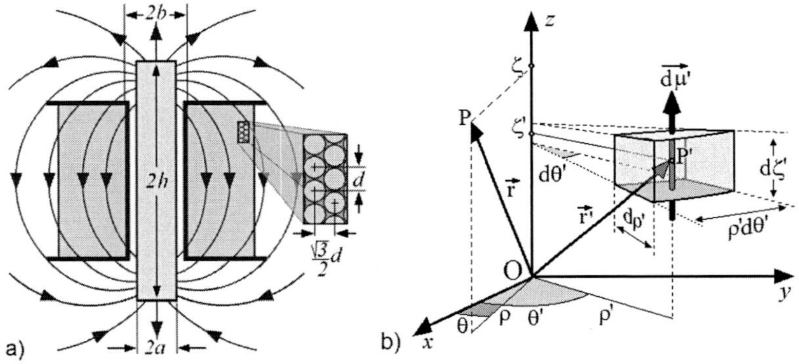

Fig. 2-42. a) The field lines generated by a uniformly magnetized bar intersecting the turns of a pick-up coil and b) a sketch of the scalar potential calculus in point *P* outside the bar.

The total potential in this point is the cumulative contribution:

$$V_m(x,y,z)=\frac{\mu_0 M_{bar}}{4\pi}\iiint\limits_{(bar)}\frac{z-z'}{\left|(x-x')^2+(y-y)^2+(z-z')^2\right|^{\frac{3}{2}}}\mathrm{d}x'\mathrm{d}y'\mathrm{d}z'$$

(2.65)

from dipoles of the whole bar volume. Due to the symmetry of the field source (the bar), it is convenient to express the scalar potential in cylindrical coordinates:

$$\rho=\sqrt{x^2+y^2}\,,\;\theta=\arctan(y/x)\,,\;\zeta=z$$

(2.66)

As far as the integration result should not explicitly depend on Φ for reasons of symmetry, we may set $\Phi = 0$ for convenience. Thus, integration over ξ' yields:

$$V_m(\rho,\zeta)=\frac{\mu_0 M_{bar}}{4\pi}\int_0^a \mathrm{d}\rho'\int_0^{2\pi}\left(\frac{1}{\sqrt{\rho^2-2\rho\rho'\cos\theta'+(\zeta-h)^2}}-\frac{1}{\sqrt{\rho^2-2\rho\rho'\cos\theta'+(\zeta+h)^2}}\right)\mathrm{d}\theta'$$

(2.67)

and the longitudinal component (normal to the coil turns) of the stay field will give the result:

$$H_\zeta(\rho,\zeta)=-\frac{1}{\mu_0}\frac{\partial}{\partial\zeta}V_m(\rho,\zeta)$$

(2.68)

The pick-up coil, assumed to be uniformly wound, contains n' layers, each of which is n'' turns (clearly, $n'\,n''=n_2$). The successive turns in a layer are separated by distance d, while the successive layers are separated by $3^{1/}2\ d/2$. Thus, the stray flux sensed by the jth turn from the ith layer will be:

$$\Phi_{stray}(\rho_i,\zeta_j)=2\pi\mu_0\int_a^{\rho_i}H_\zeta(\rho,\zeta_j)\rho\,\mathrm{d}\rho \quad \text{with} \quad \rho_i=b+\frac{\sqrt{3}d}{2}i \quad \text{and} \quad \zeta_j=\frac{2j-(n''+1)}{2}d$$

(2.69)

Summation over all the turns of the coil yields:

$$\Phi_{stray}(n',n'')=\sum_{i=1}^{n'}\sum_{j=1}^{n''}\Phi_{stray}(\rho_i,\zeta_j)$$

(2.70)

while definition (2.34), in which the intrinsic flux of the bar is $\Phi_{in}=\pi a^2\mu_0 M_{bar}$, results in:

$$q_{str}=-\frac{2}{a^2 n' n'' M_{bar}}\sum_{i=1}^{n'}\sum_{j=1}^{n''}\int_a^{\rho_i}H_\zeta(\rho,\zeta_j)\rho\,\mathrm{d}\rho$$

(2.71)

which may be evaluated via computer-aided numerical calculations.

Annex 2.8.2.

Details regarding the experimental determination of the coils and electronic circuit parameters, calibration, and several examples of $\Delta u_M(t)$ transition-type fitting functions (centered) are listed below.

- The ac current used in k_H determination was supplied by a BOP 50-2M (KEPCO) power operational amplifier, driven by a VERSATESTER (IEMI) sine wave generator and previously synchronized to the line frequency (50 Hz) to avoid beating. By means of two 6 digit precision HM-8112 (HAMEG) multimeters, one working as an AC ampermeter of 0.1 mA resolution and the other as an AC milivoltmeter of 1 μV resolution, point-by-point measurements were performed on the *FC* driving current and on the emf induced in the probe coil at different amplitudes of the current. Then, the resultant k_H was given by linear fitting of the recorded signals.
- For the determination of L_{FC}, C_{FC}, and R_{FC}, the frequency dependence of the *FC* impedance modulus was recorded by means of an HM-8118 (RHODE & SCHWARZ) LCR-bridge. The frequency response of the *FC* circuit was examined using the same BOP 50-2M -(KEPCO) VERSATESTER (IEMI) tandem, now working as a power function generator, by means of which the *ATR* was driven with a sine wave voltage of constant amplitude.
- The numbers of turns of the three sensing coils were determined with a ±1 error of the winding machine counter in the course of manufacturing.
- The frequency dependence of the impedance modulus and that of the equivalent serial inductance of the sensing coils c_1, c_2, and c_3, were recorded by means of an E-498-DA RCL meter (AGILENT). From these data, the electrical parameters and resonance frequency were determined (see the last four columns in table 1).
- For the frequency-dependent gain of the LF-357 (NATIONAL) op amp determination, a KH 5700 (KROHN-HITE) sine wave oscillator, two 6½ digit precision HM-8112 (HAMEG) multimeters working at 300 kHz bandwidth and 0.15 % full scale accuracy AC voltmeters were used. The frequency-dependent phase shift of the output signal was measured by means of a 150

MHz bandwidth MH 1508-2 (HAMEG) digital oscilloscope with on-screen reading facility for frequency and delay time.

- Using a HM-8112 (HAMEG) multimeter set to the Ohmmeter function with a resolution better than 10 Ω in the 10 kΩ range, and an E-498-DA RCL meter (AGILENT) working as a capacimeter withh resolution better than 1 nF in the 10 μF range, τ_M was evaluated to 6.287 ms. For C_f values in the μF range (a conventional upper limit), the expected R_f values are in the MΩ range, suggesting that, as far as the op amp-based inverting circuits are concerned (a category to which our integrators belong), the usual values of R_{in} are in the kΩ range, and the lower limit of the integration time $\tau_M = R_{in}C_f$ should be in the range of milliseconds at a 50 Hz working frequency. Accordingly, convenient τ_f values should be set in the range of tens of seconds.
- Using a calibrated-eyepiece microscope, the inner diameter of the glass tube containing the Fe_3O_4-based ferrofluid was measured, while the height and the mass of the Co ferrite powder column were determined using a 0.01 mm precision IP67 (ALPA) digital caliper and a 0.001 g precision SLF303 (FISHER) analytical balance.
- Examples of possible functions $\upsilon(\xi)$ used for fitting the $\Delta u_\mathrm{M}(t)$ signals:

$$\mathcal{L}(\xi)=\coth(\xi)-\frac{1}{\xi} \quad (\textit{Langevin})$$

$$\mathcal{T}(\xi)=\tanh(\xi) \quad (\textit{hyperbolic tangent})$$

$$\tilde{\mathcal{S}}(\xi)=1-2\left[1+\exp\left(\frac{\xi-a}{b}\right)\right]^{-c} \quad (\textit{asymmetric sigmoid})$$

$$\lambda(\xi)=\frac{2}{\pi}\arctan\left(\frac{\xi}{a}\right) \quad (\textit{Lorentz cumulative})$$

$$\mathcal{D}(\xi)=\frac{b}{a}\left\{\ln\left[\exp\left(\frac{c\xi+a}{b}\right)+1\right]+\ln\left[\exp\left(\frac{c\xi-a}{b}\right)+1\right]\right\}-1 \quad (\textit{cumulative double sigmoid}) \tag{2.72}$$

Annex 2.8.3.

The classical expansion of the cosine function to odd powers into a sum of cosines of odd multiple angles is given below.

$$\cos^{2m+1}(x)=2^{-2m}\left\{\cos\left[(2m+1)x\right]+C_{2m+1}^{1}\cos\left[(2m-1)x\right]+...+C_{2m+1}^{j}\cos\left[(2m-2j+1)x\right]+...+C_{2m+1}^{m}\cos(x)\right\}$$

(2.73)

It may be rewritten in the more convenient form [123]:

$$\cos^{2m+1}(x)=2^{-2m}(2m+1)!\sum_{j=0}^{m}\frac{1}{(m-j)!(m+j+1)!}\cos\left[(2j+1)x\right]$$

(2.74)

Since eq. (2.8.52) is of the form:

$$e_{mv}(t)=\sum_{j=1}^{j_{max}}e_{mv,j}(t) \qquad \text{with} \qquad e_{mv,j}(t)=-E_j\cos^{2N_j+1}\left[\omega(t-\delta t_j)\right]$$

(2.75)

using expansion (2.75), and keeping in mind that $\omega_k=(2k+1)\omega$, this may be rewritten as:

$$e_{mv,j}(t)=-\sum_{k=0}^{N_j}E_{j,k}\cos(\omega_k t-\gamma_{kj}) \qquad \textit{with} \qquad E_{j,k}=\frac{(2j+1)!}{2^{2N_j}(N_j-k)!(N_j+k+1)!}E_j$$

(2.76)

and then split to give:

$$e_{mv,j}(t)=-\sum_{k_1=0}^{N_1}E_{k_1,j}\cos\left(\omega_{k_1}t-\gamma_{k_1 j}\right)-\sum_{k_2=N_1+1}^{N_2}E_{k_2,j}\cos\left(\omega_{k_2}t-\gamma_{k_2,j}\right)..-\sum_{k_j=N_{j-1}+1}^{N_j}E_{k_j,j}\cos\left(\omega_{k_j}t-\gamma_{k_j j}\right)$$

(2.76)

If this expansion is applied to all the terms in (2.73), it is easy to see that the frequencies indexed k_1 occur in all these terms; that the frequencies indexed k_2 occur in all terms except for the first one; and that the frequencies indexed k_3 occur in all except for the first two, and so on. If the order of addends is conveniently changed by grouping the terms containing a given frequency to a single term, and with the following trigonometric identity (derived from the general form of the "Harmonic Addition Theorem" [123]):

$$\sum_i a_i \cos(x-\alpha_i)=a\cos(x-\alpha)$$

with

$$a=\sqrt{\sum_i \sum_{j\neq i} a_i a_j \cos(\alpha_i-\alpha_j)} \quad \text{and} \quad \alpha=\arctan\left(\frac{\sum_i a_i \sin\alpha_i}{\sum_i a_i \cos\alpha_i}\right)$$

(2.77)
(2.78)

is separately applied to each of these resulting terms, the form (2.56) of the test signal $e_{mv}(t)$ is the result.

2.9. Conclusion

In this chapter we have presented some methods of structural and morphological analysis of oxide materials, such as X-ray diffraction (XRD), scanning electron microscopy/spectroscopy energy-dispersive X-ray (SEM/EDX), UV-Vis spectroscopy, Fourier transform infrared spectroscopy (FTIR), and thermogravimetric analysis (TGA). After describing each method, a series of experimental data obtained for different oxide samples using different methods of preparation were presented.

The method of complex impedance was presented, which is the basis for determining the dynamic electrical properties and thermoelectrical analysis of the electrical conductivity of the investigated samples. We then described the experimental setup in the laboratory with which we measured the frequency and temperature dependence of the real and imaginary components of the complex impedance of samples, along with some of our experimental results.

The inductive method of hysteresisgraph was described, which is used for the determination of the magnetic properties of samples, along with some of our experimental results obtained using this method.

Based on complex impedance measurements, we determined the electrical conductivity of a manganese ferrite ($MnFe_2O_4$) sample in the frequency range of 100 Hz-2 MHz and temperature range of 30-130 °C.

The results show that, at all investigated temperatures, the conductivity spectrum $\sigma(f,T)$ for the investigated sample consists of both a static component σ_{DC}, corresponding to low frequencies, and an alternative component σ_{AC} that increases rapidly with frequency, corresponding to high frequencies.

From the temperature dependence of the static conductivity and based on Mott's variable-range hopping model (VRH), we determined the activation energy of the conduction processes in the investigated samples. The results show that the slope of conductivity changes at temperatures of around 70-80 °C, suggesting a possible structural transformation in the sample and also that the activation energy for conduction depends on temperature. The values obtained for the conduction activation energy are similar to those obtained for the activation energy corresponding to electrical relaxation.

The results obtained for electrical conductivity in the variable field σ_{AC} of the sample show that σ_{AC} increases abruptly at high frequencies (f> 200 BHz) and is depended on temperature. This may be due to the fact that, as the frequency increases, the ferrite nanoparticles become more active in electrical conduction due to the hopping of electrons between the Fe^{2+} and Fe^{3+} ions from the octahedral interstices, and conductivity increases with increasing frequency. These results can be explained by the theoretical *correlated-barrier hopping* model (CBH), as the conduction phenomenon in the $MnFe_2O_4$ sample is due to the hopping of charge carriers in the material between the nearest neighboring states. Using this model, we determined the energy of the bandwidth W_m of the investigated sample.

References

[1] H. Zhao, F. Pan, Y. Li. A review on the effects of TiO_2 surface point defects on CO_2 photoreduction with H2O. *J. Materiomics* 3, 1 (2017) 17–32.

[2] K. M. Ok et al. Effect of point and planar defects on thermal conductivity of TiO22x. *J. Am. Ceram. Soc.* 101, 1 (2018) 334–346.

[3] J. L. G. Fierro. Metal Oxides: Chemistry and Applications. CRC Press (2005).

[4] N. A. Abd Samad, C. W. Lai, S. B. Abd Hamid. Influence applied potential on the formation of self-organized ZnO nanorod film and its photoelectrochemical response. *Int. J. Photoenergy* (2016).

[5] G. S. Alghamdi, A. Z. Alzahrani. Bonding formation and orbitals nature of ZnO structure, *Middle East J. Sci. Res.* 13, 9 (2013) 1144–1149.

[6] G. V. Aguilar et al. Photoluminescence studies on ZnO thin films obtained by sol-gel method. In *Recent Applications in Sol-Gel Synthesis* (2017) 195.

[7] B. Ludi, M. Niederberger. Zinc oxide nanoparticles: chemical mechanisms and classical and non-classical crystallization. *Dalton Trans.* 42, 35 (2013) 12554–12568.

[8] C. Woll. The chemistry and physics of zinc oxide surfaces. *Prog. Surf. Sci.* 82, 2-3 (2007) 55–120.

[9] H. Essabir, R. Boujmal, M. O. Bensalah, D. Rodrigue, R. Bouhfid, A. E. K Qaiss. Mechanical and thermal properties of hybrid composites: oil-palm fiber/clay reinforced high density polyethylene. *Mech. Mater.* 98 (2016) 36–43.

[10] M. Raji, H. Essabir, R. Bouhfid, A. E. K. Qaiss. Impact of chemical treatment and the manufacturing process on mechanical, thermal, and rheological properties of natural fiber-based composites. In *Handbook of Composites from Renewable Materials*. Hoboken, NJ, USA: John Wiley & Sons (2017) 225–52.

[11] H. Essabir, M. Raji, E. M. Essassi, D. Rodrigue, R. Bouhfid, A. el K Qaiss. Morphological, thermal, mechanical, electrical and magnetic properties of ABS/PA6/ SBR blends with Fe_3O_4 nanoparticles. *J. Mater. Sci.-Mater. Electron.* 28, 22 (2017) 17120–30.

[12] P. Sfirloaga, I. Miron, I. Malaescu, C. N. Marin, C. Ianasi, P. Vlazan. Structural and physical properties of undoped and Ag-doped NaTaO3 synthesized at low temperatures. *Materials Science in Semiconductor Processing* 39 (2015) 721–725.

[13] I. Malaescu, A. Lungu, C. N. Marin, P. Vlazan, P. Sfirloaga, G. M. Turi, Experimental investigations of the structural transformations induced by the heat treatment in manganese ferrite synthesized by ultrasonic assisted co-precipitation method. *Ceramics International* 42, 15 (2016) 16744–16748.

[14] P. Sfirloaga, M. Poienar, I. Malaescu, A. Lungu, P. Vlazan. Perovskite type lanthanum manganite: Morpho-structural analysis and electrical investigations. *Journal of Rare Earths* 36, 5 (2018) 499–504, doi:10.1016/j.jre.2017.10.009

[15] Paula Sfirloaga, Gabriela Vlase, T. Vlase, I. Malaescu, C. N. Marin, Paulina Vlazan. Silver doping in lanthanum manganite materials: structural and electrical properties. *Journal of Thermal Analysis and Calorimetry* 142 (2020) 1817–1823.

[16] A. Lungu, I. Malaescu, C. N. Marin, P. Vlăzan, P. Sfirloaga. The electrical properties of manganese ferrite powders prepared by two different methods. *Physica B: Condensed Matter* 462 (2015) 80–85.
[17] I. Malaescu, A. Lungu, C. N. Marin, P. Sfirloaga, P. Vlazan, S. Brindusoiu, M. Poienar. Temperature dependence of the dynamic electrical properties of Cu1+xMn1-xO2 (x = 0 and 0.06) crednerite materials. *Ceramics International* 44 (10) (2018) 11610–11616, doi: 10.1016/j.ceramint.2018.03.229.
[18] S. Brindusoiu, M. Poienar, C.N. Marin, P. Sfirloaga, P. Vlazan, I. Malaescu. The electrical conductivity of $Fe_3(PO_4)_2 \cdot 8H_2O$ materials. *Journal of Materials Science: Materials in Electronics* 30, 16 (2019) 15693–15699.
[19] I. Mihalca, A. Ercuta, C. Ionascu. *Sensors and Actuators A: Physical* 106, 1-3 (2003) 61–64.
[20] Jens Als-Nielsen and Des McMorrow. *Elements of Modern X-ray Physics*. John Wiley & Sons (2001), ISBN-10:0471498580.
[21] D. Keith Bowen, Brian K. Tanner. *High Resolution X-ray Diffractometry and Topography*. Taylor & Francis (1998), ISBN-10:0850667585.
[22] D. Bally, L. Benes, R. Manaila, Ed. Tehn. Bucuresti (1972).
[23] Handbook of Chemistry and Physics, 52nd ed. (1971-72). The Chemical Rubber Company, Cleveland, Ohio, USA.
[24] H. M. Rietveld. A profile refinement method for nuclear and magnetic structures. *J. Appl. Cryst.* 2 (1969) 65–71.
[25] G. Malmros, J. O. Thomas. Least-squares structure refinement based on profile analysis of powder film intensity data measured on an automatic microdensitometer. *J. Appl. Cryst.* 10 (1977) 7–11.
[26] C. P. Khattak and D. E. Cox. Profile analysis of X-ray powder diffractometer data: structural refinement of La0.75Sr0.25CrO3. *J. Appl. Cryst.* 10 (1977) 405.
[27] P. Sfirloaga, I. Malaescu, M. Poienar, M.C. Nicolae, D. Malaescu, P. Vlazan. Synthesis, structural and electrical properties of NaTaO3:Cu. *J. Mater. Sci.: Mater. Electron.* vol. 27, 11 (2016) 11640–11645.
[28] M. Stoia, C. Pacurariu, C. Mihali, I. Malaescu, C. N. Marin, A. Capraru. Manganese ferrite-polyaniline hybrid materials: Electrical and magnetic properties. *Ceramics International* 45, 2 (2019) 2725–2735, doi.org/10.1007/s10973-020-10095-1.
[29] P. Sfirloaga, M. Poienar, I. Malaescu, A. Lungu, C. V. Mihali, P. Vlazan. Electrical conductivity of Ca-substituted lanthanum manganites. *Ceramics International* 44, 6 (2018) 5823–5828, doi:10.1016/j.ceramint.2018.01.029.
[30] P. Sfirloaga, I. Malaescu, C. N. Marin, P. Vlazan, The effect of partial substitution of Pd in $LaMnO_3$ polycrystalline materials synthesized by sol-gel technique on the electrical performance. *Journal of Sol-gel Science and Technology* 92, 3 (2019) 537 *Ceramics International* 545, doi:10.1007/s10971-019-05102-3.
[31] Amighian, J., Mozaffer, M., and Nasr, B. Preparation of nano-sized manganese ferrite (MnFe2O4 via coprecipitation method). *Physica Status Solidi. C. Current Topics in Solid State Physics* 3 (2006) 3188–3192.

[32] Joint Committeeon Powder Diffraction Standards: International Center for Diffraction Data, *Swarthmore* (1993).
[33] Nasser Y. Mostafa, M. M. Hessiena, Abdallah A. Shaltouta. Hydrothermal synthesis and characterizations of Ti substituted Mn-ferrites. *Journal of Alloys and Compounds* 529 (2012) 29–33.
[34] H. P. Klug, L. E. Alexander. X-Ray Diffraction Procedures. (2nd ed.) John Wiley & Sons, New York (1974).
[35] Y. M. Z. Ahmed. Synthesis of manganese ferrite from non-standard raw materials using ceramic technique. *Ceramics International* 36 (2010) 969–977.
[36] P. K. Panda, B. Sahoo. PZT to Lead Free Piezo Ceramics: A Review. *Ferroelectrics* 474 (2015) 128.
[37] J. Rodel, K. G. Webber, R. Dittmer, W. Jo, M. Kimura, D. Damjanovic. Transferring Lead-free Piezoelectric Ceramics into Application. *J. Eur. Ceram. Soc.* 35 (2015) 1659.
[38] M. Valant, A.-K. Axelsson, N. Alford. Review of Ag(Nb, Ta)O_3 as a Functional Material. *J. Eur. Ceram. Soc.* 27 (2007) 2549.
[39] W. Wunderlich, $NaTaO_3$ composite ceramics: A new thermoelectric material for energy generation. *J. Nuclear Mater.* 389 (2009) 57–61.
[40] P. Sfirloaga, C. N. Marin, I. Malaescu, P. Vlazan. The electrical performance of ceramic mateials with perovskite structure doped with metallic ions. *Ceramics International* vol. 42, 16 (2016) 18960–18964.
[41] H. Kato, K. Asakura, A. Kudo. *J. Am. Chem. Soc.* 125 (2003) 3082.
[42] M. Yoshino, M. Kakihana,W. S. Cho, H. Kato, A. Kudo. *Chem. Mater.* 14 (2002) 3369–3376.
[43] W. H. Lin, C. Cheng, C. C. Hu, H. S. Teng. *Appl. Phys. Lett.* 89 (2006) 211904.
[44] J. I. Goldstein, D. E. Newbury, P. Echin, D. C. Joy, C. E. Lyman, E. Lifshin, L. Sawyer, J. R. Michael. Scanning Electron Microscopy and X-Ray. (3rd ed.) Springer Science (2003).
[45] D. Bojin, D. Bunea, F. Miculescu, *Microscopie electronică de baleiaj şi aplicaţii*. Editura Agir, Bucureşti (2005).
[46] http://en.wikipedia.org/wiki/Energy-dispersive_X-ray_spectroscopy.
[47] A. Douglas, F. Skoog, James Holler, Stanley R. Crouch. *Principles of Instrumental Analysis*. 6th ed., 173. ISBN-13: 978-0495012016 (2006).
[48] Prabhakar Misraand, Mark Dubinskii. *Ultraviolet Spectroscopy and UV Lasers* Marcel Dekker (ed.), New York (2002). ISBN: 0-8247-0668-4.
[49] C. P. Stemmet, J. C. Schouten, T. A. Nijhuis. *Chem. Eng. Sci.* 65 (2010) 267.
[50] Wan-Hsien Lin, Ching Cheng, Che-Chia Hu, Hsisheng Teng. $NaTaO_3$ photocatalysts of different crystalline structures for water splitting into H_2 and O_2. *Applied Physics Letters* 89, 21 (2006) 211904.
[51] K. Roy Sumit, S. N. Singh, K. Kumar, K. Prasad, Structural, FTIR and ac conductivity studies of NaMeO3 (Me≡Nb, Ta) ceramics. *Adv. Mater. Res.* 2 (2013) 173–180.
[52] Y.-X. Zhao, D.-R. Liu, F.-F. Li, D.-F. Yang, Y.-S. Jiang, Preparation, characterization and photocatalytic activity of N-doped NaTaO3 nanocubes. *Powder Technol.* 214 (2011) 155–160.

[53] A. Sobhani-Nasab, M. Maddahfar, S. M. Hosseinpour-Mashkani. *J. Mol. Liq*. 216, 1 (2016).

[54] O. Vázquez-Cuchillo, A. Manzo-Robledo, R. Zanella, N. Elizondo-Villareal. A. Cruz-López, *Ultrasonics Sonochemistry* 20, 498 (2013).

[55] D. Malaescu, I. Grozescu, P. Sfirloaga, P. Vlazan, C. N. Marin. The Electrical Properties of Some Composite Materials Based on Sodium and Tantalum Oxides. *Acta Physica Polonica A* vol. 129, 1 (2016) 133–137.

[56] A. Lungu, P. Vlazan, D. Malaescu, I. Malaescu, M. Lungu, P. Sfirloaga, R. Bianca, M. Poienar. Ultrasounds assisted co-precipitation syntesis and photocatalytic activity of crednerite $CuMnO_2$. *The International Conference on Powder Metallurgy and Advanced Materials*, Cluj-Napoca, RoPM-AM (2017).

[57] Nobuhito Imanaka, Toshiyuki Masui, Hidekazu Hirai, Gin-ya Adachi, Amorphous Cerium-Titanium Solid Solution Phosphate as a Novel Family of Band Gap Tunable Sunscreen Materials. *Chem. of Mater*. 15, 12 (2003) 2289–2291.

[58] Shaozheng Hua, Lin Ma, Jiguang You, Fayun Li, Zhiping Fan, Guang Lu, Dan Liu, Jianzhou Gui. Enhanced visible light photocatalytic performance of g-C_3N_4 photocatalysts co-doped with iron and phosphorus. *Applied Surface Science* 311 (2014) 164–171.

[59] Griffiths, P., de Hasseth, J. A. (18 May 2007). Fourier Transform Infrared Spectrometry (2nd ed.) Wiley-Blackwell, ISBN:978-0-471-19404-0.

[60] N. Colthup, L. Daly, S. Wiberley. Introduction to Infrared and Raman Spectroscopy (3rd ed.), Academic Press, USA (1990), ISBN:9780080917405.

[61] V. Saptari, Chapter 1: Spectroscopy Instrumentation. In *Fourier Transform Spectroscopy Instrumentation Engineering*, SPIE (2003).

[62] Stuart, B. H. Introduction. In *Infrared Spectroscopy: Fundamentals and Applications*; Ando, D. J., Stuart, B. H. (eds.) John Wiley & Sons (2004) 1–13, ISBN:978-0-470-01114-0.

[63] Z. Lazarevi¢, C. Jovaleki¢, A. Milutinovi¢, M. J. Romcevi¢, N. Romcevi¢. Preparation and Characterization of Nano Ferrites. *Acta Phys. Pol. A*. 121 (2012) 682–686.

[64] Hugh R. Carlon. Christiansen effect in IR spectra of soil-derived atmospheric dusts. *Appl. Optics* 18, 21 (1979) 3613–3614.

[65] M. J. Pollard, P. R. Griffiths, K. Nishikida. Investigation of the Christiansen Effect in the Mid-Infrared Region for Airborne Particles. *Appl. Spectr.* 61, 8 (2007) 860–866.

[66] *** Infrared and Raman Spectroscopy-Principles and Spectral Interpretation. Peter Larkin Elsevier (2011), ISBN: 978-0-12-386984-5.

[67] A. W. Coats and J. P. Redfern. Thermogravimetric Analysis: A Review. *Analyst* 88,1053 (1963) 906–924.

[68] M. Poienar, P. Sfirloaga, C. Martin, D. Ursu, P. Vlazan, Hydrothermal synthesis of crednerite CuMn12xMxO2 (M 5 Mg, Al; x 5 0–0.08): structural characterisation and magnetic properties. *J. Maer. Sci.: Ceramics* (2017).

[69] J. R. McDonald. Impedance spectroscopy. John Wiley & Sons, New York 5 (1987).

[70] Evgenij Barsoukov, J. Ross Macdonald (eds.). *Impedance Spectroscopy Theory, Experiment, and Applications*. John Wiley & Sons, Hoboken, New Jersey (2005).
[71] Suchismita Mohanty, R. N. P. Choudharyn, R. Padhee, B. N. Parida. *Ceramics International* 40 (2014) 9017–9025.
[72] Khalid Mujasam Batoo. *Physica B* 406 (2011) 382–387.
[73] B. K. P. Scaife. *Principles of Dielectrics*. Clarendon Press, Oxford (1998).
[74] S. Khadhraoui, A. Triki, S. Hcini, S. Zemni, M.Oumezzine. *J. Magn. Magn. Mater.* 371 (2014) 69–76.
[75] P. Debye. *Polar Molecules*. Dover, New York (1929).
[76] N. Siva Kumar, A. Narayanasamy, K. Shinoda, C. N. Chinnasamy, B. Jeyadevan, J. M. Greneche. *J. Appl. Phys*. 102 (2007) 013916.
[77] E. A. Davis, N. F. Mott. Conduction in non-crystalline systems V. Conductivity, optical absorption and photoconductivity in amorphous semiconductors. *Philos. Mag.* 22 (1970) 903–922.
[78] S. R. Elliott. A theory of a.c. conduction in chalcogenide glasses. *Philos. Mag*. 36 (1977) 1291–1304.
[79] M. Nadeem, M. J. Akhtar, A. Y. Khan. *Solid State Commun.* 134 (2005) 431.
[80] A. K. Jonscher. *Universal Relaxation Law*. (1st ed.) Chelsea Dielectrics Press, London (1996).
[81] C. K. Suman, K. Prasad, R. N. P. Choudhary. *J. Mater. Sci.* 41 (2006) 369–375.
[82] Dalibor L. Sekulić, Zorica Z. Lazarević, Čedomir D. Jovalekić, Aleksandra N. Milutinović, Nebojša Z. Romčević. Impedance Spectroscopy of Nanocrystalline $MgFe_2O_4$ and $MnFe_2O_4$ Ferrite Ceramics: Effect of Grain Boundaries on the Electrical Properties. *Science of Sintering* 48 (2016) 17–28.
[83] M. A. El Hiti. *Journal of Magnetism and Magnetic Materials* 164, 187 (1996).
[84] K. Funke. *Prog. Solid State Chem.* 22 (1993) 111–195.
[85] C. G. Koops. *Phys. Rev.* 83 (1951) 121–124.
[86] A. S. Fawzi, A. D. Sheikh, V. L. Mathe. *J. Alloys Compd.* 502 (2010) 231–237.
[87] N. F. Mott and E. A. Davis. *Electronic Process in Non-crystalline Materials*. Clarendon Press, Oxford (1979).
[88] N. Ponpandian, P. Balaya, A. Narayanasamy. Electrical conductivity and dielectric behaviour of nanocrystalline $NiFe_2O_4$ spinel. *J. Phys.: Condens. Matter* 14 (2002) 3221–3237.
[89] Abdeen A. M. *J. Magn. Magn. Mater.* 192 (1999) 121.
[90] Chen Ang, J. R. Jurado, Zhi Yu, M. T. Colomer, J. R. Frade, J. L. Baptista. *Phys. Rev. B* 57 (1998) 19.
[91] Hill R. M. Hopping conduction in amorphous solids. *Phil. Mag.* 24 (1971) 1307.
[92] S. R. Elliot. *Adv. Phys.* 36 (1987) 135.
[93] Fayek M. K., Mostafa M. F., Sayedahmed F., AtaAllah S. S., Kaiser M. *J. Magn. Magn. Mater.* 210 (2000) 189.

[94] J. Judith Vijaya, G. Sekaran, M. Bououdina. Effect of Cu^{2+} doping on structural, morphological, optical and magnetic properties of $MnFe_2O_4$ particles/sheets/flakes-like nanostructures. *Ceramics International* volume 41, issue 1, part A, January (2015) 15–26.
[95] M. Ustundağ, M. Aslan. Electronic and Magnetic Properties of Ca-Doped Mn-Ferrite. *Acta Phys. Pol. A* vol. 130 (2016) 362–364.
[96] S. Foner. *J. Appl. Phys.* 79 (1996) 4740–45. Hoon S. R. *Eur. J. Phys.* 4 (1983) 6147.
[97] P. J. Flanders. *Rev. Sci. Instrum.* 61 (1990) 839–47.
[98] K. Enpuku, T. Minotani, T. Gima, Y. Kuroki, Y. Itoh, M. Yamashita, Y. Katakura, S. Kuhara. *Japan J. Appl. Phys.* 38 (1999) L1106.
[99] K. Koyama, S. Hane, K. Kamishima, T. Goto, *Rev. Sci. Instrum.* 69 (1998) 3009.
[100] S. Legl, C. Pfleiderer, K. Krämer. *Rev. Sci. Instrum.*, 81, no. 4, Apr. 2010.
[101] W. Zhou, T. Seki, K, Takahashi. *J. Appl. Phys* 122 (9) (2017) 093902.
[102] R. Egan, M. Phillipe, L. Wera, J. F. Fagnard, B. Vanderheyden, A. Dennis, Y. Shi, D. A. Cardwell, P. Vanderbemden. *Rev. Sci. Instrum.* 86 (2015) 0.25107.
[103] H. J. Oguey. *Rev. Sci. Instrum.* 33 (1960) 701–09.
[104] H. Fay. *Rev. Sci.Instrum.* 43 (1972) 1274–79.
[105] A. Ercuta. *IEEE Transactions on Instrumentation and Measurement* 69 (4) (2020) 1643–1651.
[106] F. Fiorillo. *Measurement and Characterization of Magnetic Materials*. Amsterdam: Elsevier (2004) 372.
[107] E. C. Crittenden Jr, A. A. Hudimac, R. I. Strough. *Rev. Sci. Instrum.* 22 (1951) 872.
[108] S. A. Shah, R. M. Ferguson, K. M. Krishnan. *J. Appl. Phys* 116 (2014) 163910.
[109] R. Scholes. *IEEE Trans. Magn.* 6 (2) (1970) 289–91.
[110] D. X. Chen, E. Pardo, A. Sanchez. *IEEE Trans.Magn.* 38 (2002) 1742–52.
[111] S. Chikazumi. *Physics of Ferromagnetism*. New York: Oxford Univ. Press (1997) 33.
[112] E. du Tremolet de Lacheisserie, D. Gignoux, M. Schlenker (eds.). *Magnetism. Materials and Applications Vol II*. Boston: Springer Science & Business Media, Inc. (2005) 458.
[113] M. Butta, G. Infante, P. Ripka, G. A. Badini-Confalonieri, M. Vazquez. *Rev. Sci. Instrum.* 80 (2009) 083906.
[114] S. Tumanski. *Handbook of magnetic measurements*. Boca Raton: CRC Press (2011) 169
[115] M. Stefanescu, M. Bozdog, C. Muntean, O. Stefanescu T. Vlase. *J. Mag. Magn. Mater.* 393 (2015) 92–98.
[116] M. Stoia, C. Pacurariu, R. Istratie, D. Niznansky. *J. Therm. Anal. Calorim.* 121 (2015) 989–1001.
[117] C. B. Craus, A. R. Chezana, M. H. Siekmanb, J. C. Lodderb, D. O. Boerma, L. Niesen. *J. Mag. Magn. Mater.* 240 (2002) 423–426.
[118] I. Malaescu, N. Stefu, L. Gabor, *J. Mag. Magn. Mater.* 234 (2001) 299–305.

[119] A. Ercuta, M. Chirita. *J. Crystal Growth* 380 (2013) 182–186.
[120] G. Chrystal. *Algebra: an elementary textbook.* (7th ed.) part II Providence, Rhode Island: American Mathematical Society Chelsea Publ. (1964) 278.
[121] I. I. Cotaescu. *Private communication* (2012).
[122] J. Vanderlinde. *Classical electromagnetic theory.* (2nd ed.) Dordrecht: Kluwer Academic Publisher (2004) 197.
[123] E. W. Weisstein. *CRC Concise Encyclopedia of Mathematics.* (2nd ed.) Boca Raton: Chapman & Hall/CRC (2003) 1302.

CHAPTER 3

OXIDE MATERIALS WITH SPINEL STRUCTURES

PAULINA VLĂZAN, ANTOANETTA LUNGU, PAULA SFIRLOAGĂ, CATALIN N. MARIN, AUREL ERCUȚA, IOSIF MALAESCU

Spinel materials, in general, are a class of metallic oxides with the chemical formula AB_2O_4, where A represents bivalent ions (Zn, Mn, Fe, Co, Ni, or Cu, etc.) and B represents trivalent ions (Al, Ga, Fe, or Cr, etc.) or tetravalent ions (Si, Ti, Ge, or Sn, etc.). Spinel compounds crystallize in cubic space group Fd-3m and can be used as perspective-functional materials in various fields of application, including: water treatment; degradation of dyes; catalysis; photocatalysis, medicine; hyperthermia; drug delivery; and as antibacterial agents. In this chapter, spinel $MnFe_2O_4$ materials are described and, considering the complexity of these materials, the structural and physical properties are presented.

3.1. Introduction

Compounds with spinel structures are, generally, a class of metal oxides with the chemical formula AB_2O_4, where A represents bivalent ions (A^{2+}: Zn, Mn, Fe, Co, Ni, Cu, etc.) and B represents trivalent ions (B^{3+}: Al, Ga, Fe, or Cr, etc.) or tetravalent ions (B^{4+}: Si, Ti, Ge, or Sn, etc.) [1]. Spinel compounds crystallize in a cubic system with centered faces having the Fd-3m space group (no. 227), which has two types of voids: tetrahedral and octahedral. In this structure, the oxygen anions (O^{2-}) form a relatively tight cubic lattice with centered faces, in which the metal cations occupy the interstitial positions in the arrangement of the two networks (tetrahedral and octahedral) [2]. The spinel architectures are flexible in terms of the arrangement of cations in the interstices of the two lattices (tetrahedral and octahedral). Thus, they can form compounds with the normal, inverse, or mixed spinel structures [3, 4]. The face-centered cubic crystal structure has a unit cell containing 32 O^{2-}, 8 Me^{2+}, and 16 Fe^{3+} ions. The oxygen ions form 64 tetrahedral and 32 octahedral sites, with a distribution of 24 cations. The eight Me^{2+} and eight Fe^{3+} cations occupy half of the octahedral sites, while the other eight Fe^{3+} ions occupy eight tetrahedral sites [5].

Manganese ferrites ($MnFe_2O_4$) belong to this class of important magnetic materials with outstanding structural and physical properties. It is a soft ferrite with a mixed cubic spinel structure and the bivalent (Mn^{2+}) and trivalent (Fe^{3+}) ions occupy both tetrahedral (A) and octahedral (B) sites [6]. The properties of these materials are influenced by several factors including: the method of preparation; chemical composition; and conditions of synthesis (reaction time, temperature, and reagent concentrations). For processes that take place in solution, the pH plays an important role in obtaining the spinel phase of $MnFe_2O_4$. Another important factor that influences the properties of ferritic materials is heat treatment (temperature, along with time and rate of temperature increase). In addition, ferrite properties depend on the crystallite size, morphology, the nature and concentration of the dopant, and the distribution of cations between the tetrahedral (A) and octahedral (B) sites in the spinel MFe_2O_4 structure [7]. It is known that the properties of nanomaterials are different to those of the corresponding bulk materials.

There are a number of methods widely used for the preparation of ferrites, such as the *co-precipitation* method, which is an economical and efficient method for the synthesis of ultrafine ferrite powders [8]; the *hydrothermal method*, which has several advantages compared to conventional processes for maintaining control over the composition, shape, particle size and distribution, and low temperature processing in the presence of a proper solvent etc. [9, 10]; the *sol-gel* method [11, 12]; the combustion method [13]; and the microemulsion method [14].

Spinel ferrite nanomaterials have been extensively studied in recent decades, due to their low costs and widespread potential applications in telecomunications, electronic devices, magnetic sensors, recording equipment, magnetic fluids, magnetic resonance imaging (MRI), and microwave absorbers, etc. Ferrite nanomaterials have many uses in different fields, including medicine, energy, and environmental protection. In the biomedical field, they have applications such as targeted drug delivery [15], as contrast agents in magnetic resonance imaging (MRI) [16], and in clinical *cancer treatment* [17].

Applications for environmental protection involve the removal of heavy metal ions and organic dyes, which have harmful impacts on living organisms [18-20]; microwave absorbers [21, 22]; and energy production and storage [23]. These applications are based upon the very basic properties of such materials, such as significant saturation magnetization, high electrical resistivity, low electrical losses, and very good chemical stability. The doping with transition metals ions can induce significant structural distortion and changes to the optical and magnetic properties of spinel ferrites by partial substitution of Fe^{3+} ions [24]. In this book, we present our studies into the effect of doping of Ni/Zn on physicochemical properties of the manganese ferrite nanopowders obtained by the hydrothermal method. The introduction of different metal ions into ferrite materials with spinel structures leads to changed physicochemical properties.

3.2. Experiments

3.2.1. Sample preparation of manganese ferrites by the hydrothermal method

To prepare $MnFe_2O_4$ nanomaterials by the low temperature hydrothermal process, the following precursors were used: 1 mmol iron nitrate ($Fe(NO_3)_3{\cdot}H_2O$) and 0.5 mmol manganese nitrate ($Mn(NO_3)_2{\cdot}4H_2O$). For doping of the manganese ferrite, 1 % nitrate of nickel ($Ni(NO_3)_2{\cdot}6H_2O$) or zinc ($Zn(NO_3)_2{\cdot}6H_2O$) were used as precursors. Stoichiometric amounts of these reagents were dissolved in 50 ml of deionized water. The mixtures were stirred with a magnetic stirrer until the reactants were completely dissolved. Then, 2 M NaOH solution was added until pH = 12 for precipitation of the ferrite phases. The temperature was increased to 50 °C and this was maintained for 30 minutes to complete the reaction. The mixtures were then transferred into Teflon-lined stainless steel autoclaves with 50 ml capacity. The autoclaves were sealed and placed in an oven at 200 °C. This temperature was maintained for 12 hours, before the mixtures were allowed to cool down to room temperature. The precipitates obtained were separated by settling and filtration, before being washed several times by deionized water and ethanol to remove unwanted products. The samples were finally dried in an oven at 80 °C for 4 h.

The obtained samples were characterized by powder X-ray diffraction (XRD), Fourier transform-infrared (FT-IR) spectroscopy, thermal analysis (TG-DTA), scanning electron microscopy (SEM), energy dispersive X-ray (EDX) spectroscopy, and UV-Vis diffuse reflectance spectroscopy (DRS). Making complex impedance measurements in the frequency range 100 Hz-2 MHz at different temperatures between 30 and 130 °C, we performed a comparative experimental study on the frequency and temperature dependence of the complex impedance of the samples, the effect of Ni or Zn ion doping on the static and alternating current electrical conductivity of the investigated samples. We compared the obtained results with an appropriate theoretical model described in the specialized literature. At the same time, we investigated the magnetic properties of the samples.

3.2.2. Characterization techniques

Structural and phase analysis of the prepared samples was performed by X-ray diffraction (XRD) using a PANalytical X'Pert PRO MPD diffractometer with Cu Kα radiation at $\lambda = 1.5406$ Å and a 2θ-step of 0.01° from 20 to 80°. The Fourier transform infrared spectroscopic analysis was recorded by a Shimadzu Prestige FT-IR spectrometer with a scanning range of 400-4000 cm^{-1}. Surface morphological studies and energy dispersive X-ray analysis of the samples were performed with an Inspect S scanning electron microscope (SEM) equipped with electron dispersive X-ray (EDX) for elemental chemical analysis. The UV-Vis diffuse reflectance spectrum (DRS) was recorded using UV/Vis/NIR Perkin Elmer Lambda 950 spectrophotometer. The thermogravimetric and differential thermal analysis were performed using simultaneous TG-DTA apparatus thermal analyzer LABSYS evo from SETARAM.

3.3. Results and Discussion

3.3.1. Characterization of nanomaterials

3.3.1.1. X-ray diffraction

Figure 3-1 a) shows the XRD pattern of undoped and Ni/Zn doped $MnFe_2O_4$ powders prepared by the hydrothermal method. The structure and phase purity for the obtained samples were studied by analysing the X-ray diffraction (XRD) patterns of the powder. The observed diffraction peaks can be indexed as (220), (311), (222), (400), (422), (511), (440), (530), (620), (533), (622), and (444) planes, which are characteristic of a single phase spinel structure according to JCPDS card no. 01-074-2403. Figure 3-1b) indicates that the main peaks (311) of the diffraction spectra are slightly shifted right for Ni/Zn doped samples, suggesting that the crystal lattice of $MnFe_2O_4$ changes with the nature of the impurity and depending on its ionic radius.

The X-ray diffraction (XRD) spectra for undoped and Ni/Zn doped $MnFe_2O_4$ samples were analyzed using the X'Pert HighScore Plus tool by use of the Rietveld refinement technique. The XRD patterns for all the samples were refined using the Fd-3m space group. The values of the lattice parameters and crystallites sizes obtained from the Rietveld analysis are reported in table 3-1.

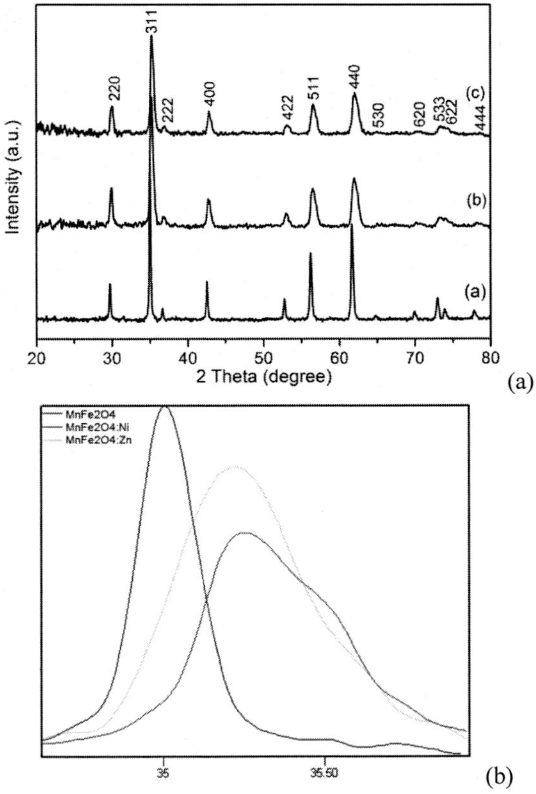

Fig. 3-1. a) X-ray diffraction for undoped $MnFe_2O_4$ and doped Ni/Zn obtained by the hydrothermal method at 200 °C for 12 h and b) the Bragg peaks (311) are shifted to the right.

The Rietveld refinement data indicate that, the inclusion of an Ni^{2+} ion in the $MnFe_2O_4$ crystal lattice leads to a slight increase in the lattice parameter, while in the case of a Zn^{2+} ion, a decrease in the lattice parameter results. This can be attributed to the fact that Ni^{2+} and Zn^{2+} ions have diferent ionic radii to those of Mn^{2+} and Fe^{3+}.

Table 3-1.

Sample	Lattice parameters a=b=c	Space group $\alpha=\beta=\gamma=90°$	Unit Cell Volume [Å^3]	Crystallite size [nm]
$MnFe_2O_4$	8.498(1)	Fd-3m	613.667	45.0
$MnFe_2O_4$:Ni	8.510(2)	Fd-3m	615.874	27.4
$MnFe_2O_4$:Zn	8.480(1)	Fd-3m	610.353	23.9

- The average crystallite sizes of undoped and Ni/Zn doped $MnFe_2O_4$ nanomaterials were calculated using the Debye-Scherrer equation [25]:

$$D_{hkl} = K\lambda/(\beta_{hkl}\cos\theta)$$

where:

- D_{hkl} is the crystallite size perpendicular to the lattice planes;
- hkl represents the Miller indices of the analyzed planes;
- K is a dimensionless shape factor with a value close to unity, which varies with the shape of the crystallite;
- λ is the wavelength incident radiation of x-ray diffraction;
- β is the width at half the maximum intensity (FWHM) of the X-ray diffraction peak for Cu Kα radiation (λ = 0.15406 nm);
- and θ is the Bragg diffraction angle.

From the results presented in table 3-1, we can observe that the crystallite size for undoped $MnFe_2O_4$ nanoparticles is higher than the average crystallite size for Ni/Zn-doped $MnFe_2O_4$.

3.3.1.2 Fourier transform infrared spectroscopy (FTIR) analysis

The FTIR spectra for the samples prepared by hydrothermal method are shown in fig. 3-2.

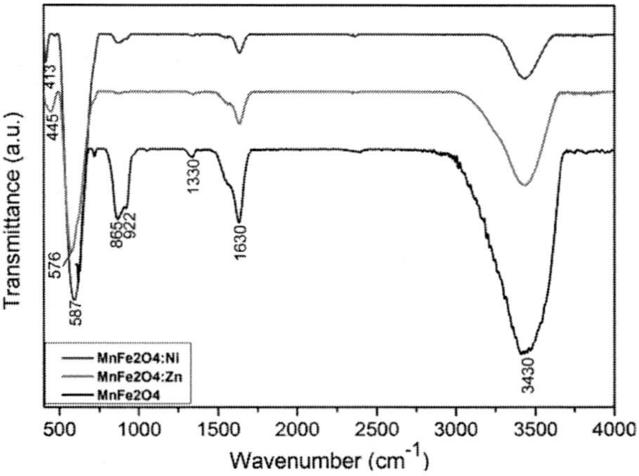

Fig. 3-2. FT-IR spectra for undoped and Ni/Zn doped $MnFe_2O_4$.

The FTIR spectra of the prepared samples shows characteristic absorption bands around 413-587 cm^{-1}, which is due to the bond vibrations of metal-oxygen complexes in the tetrahedral and octahedral sites of the spinel. As such, the FT-IR spectra show low intensity bands in the range of 413-445 cm^{-1}, which can be attributed to the Mn–O stretching vibration band in $MnFe_2O_4$. The intense peaks characteristic of metal oxides from 576 and 587 cm^{-1} can be assigned to metal-oxygen bonds (Mn-O/Fe-O) in the manganese ferrite [26]. The different frequencies of these vibrations have been attributed to differences in bond length of the metal-oxygen for tetrahedral and octahedral sites: the bond length of metal-oxygen in tetrahedral sites is smaller than the bond length of metal-oxygen in octahedral sites and has an inverse relationship with the absorption band [27, 28]. The broad band vibration from 3430 cm^{-1} can be associated with the O–H stretching vibration of the adsorbed water molecules indicating higher amounts of surface OH. The absorption band located at 1630 cm^{-1} can be ascribed to the stretching and bending modes of water adsorbed on the surface of the nanoparticles and hydroxyl (–OH) groups, which are retained during preparation, because the samples were prepared in aqueous solution by the hydrothermal route [29, 30].

3.3.1.3. Thermal analysis

The thermal decomposition of $MnFe_2O_4$ powders was analyzed by TG/TGA with the LABSYS evo apparatus by heating up to 800 °C with a heating rate of 10 °C/min. The measurements were carried out in an air atmosphere (42.8 ml/min) with 24 mg of powder in alumina crucibles.

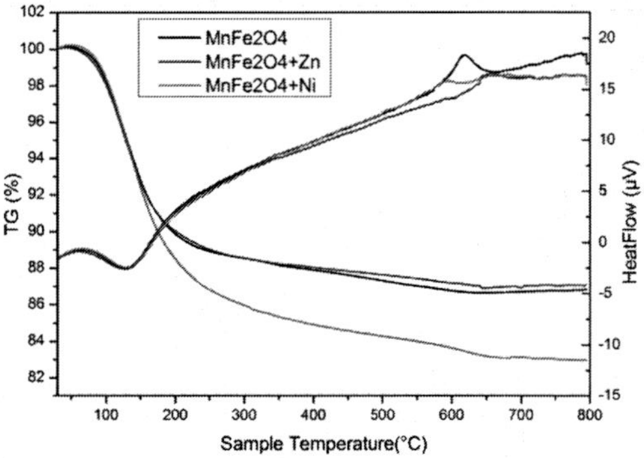

Fig. 3-3. TG and DTA curve for undoped and Ni/Zn doped $MnFe_2O_4$.

The TGA curve (fig. 3-3) indicates that from 100 to 235 °C there is a weight loss of about 12-15 %, which can be attributed to the dehydration of the amorphous powder and its transition to the crystalline phase. The DTA curves shows endothermic peaks at around 580 to 680 °C for synthesized pure $MnFe_2O_4$ powders and Ni/Zn doped powders, which is due to the ferrite decomposition reaction. At higher temperatures, $MnFe_2O_4$ is unstable in air and Mn^{2+} ions oxidize to convert into Mn^{3+}, which results in the formation of other phases along with $MnFe_2O_4$. Such reactions would depend on the partial pressure of oxygen in the environment and the heat treatment temperature [31].

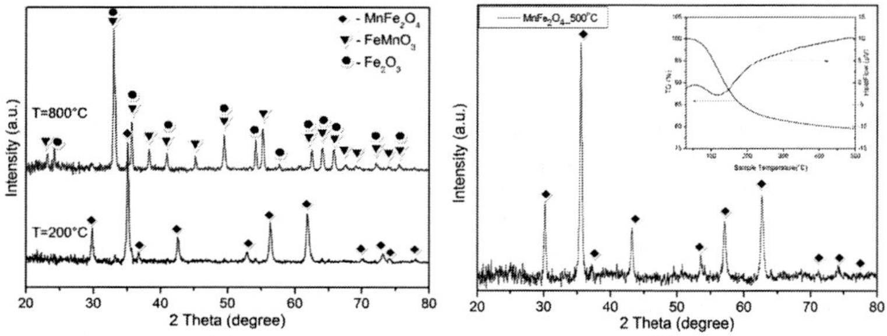

(a) (b)

Fig. 3-4. XRD patterns of annealed powder at 500 and 800 °C in air for undoped and Ni/Zn doped $MnFe_2O_4$.

As such, in fig. 3-4 a) we can observe that when the annealing temperature is higher than 550 °C, we see the transformation of the $MnFe_2O_4$ spinel phase in the $MnFeO_3$ perovskite phase and Fe_2O_3. However, ferrite powders that have been thermally treated up to 500 °C exhibit a single spinel phase (fig. 3-4 b)).

3.3.1.4 Morphological analysis

The morphology and size of the nanoparticles prepared by the hydrothermal method were studied through scanning electron microscopy (SEM). The SEM images for undoped and Ni/Zn doped $MnFe_2O_4$ samples prepared by the hydrothermal method, are shown in fig. 3.5. The images indicate a uniform distribution of spherically-shaped particles, agglomerated in clusters due to the presence of the magnetic interactions between them. Microstructural analysis gives information about the average size and type of particle morphology, which influences the properties of materials.

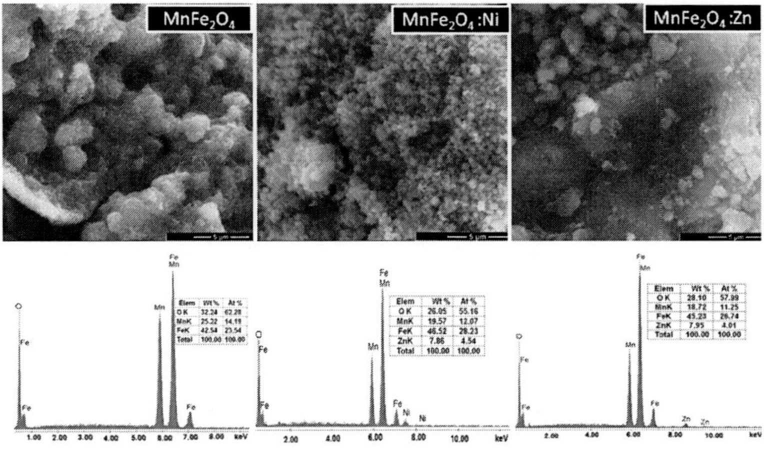

Fig. 3-5. SEM-EDX for undoped and Ni/Zn doped $MnFe_2O_4$.

Quantification of the EDX spectrum showed the elementary composition of the samples. $MnFe_2O_4$ doped with Ni/Zn was shown to have Ni content of up to 7.86 % wt% and Zn 7.95 % wt%, which are both close to the expected theoretical values.

3.3.1.5. UV-Vis analysis

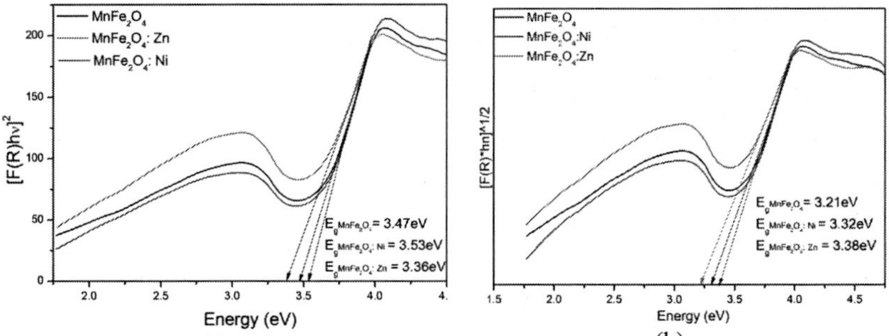

(a) (b)

Fig. 3-6. Direct (a) and indirect (b) band gap spectral determination using the Kubelka-Munk equation for undoped and Ni /Zn doped $MnFe_2O_4$ materials.

UV-Vis diffuse reflectance analysis was used to study the effect of Ni/Zn doping of $MnFe_2O_4$ on its optical properties. Diffuse reflectance has advantages over absorption spectroscopy for studying the optical properties of the powder samples with decreased scattering [32]. This was done to observe the effect of doping with Ni/Zn on the optical properties of pure spinel $MnFe_2O_4$. The diffuse reflectance data were used to calculate the direct band gap energy (Eg) using the Kubelka-Munk function F(R). The Kubelka-Munk function is directly proportional to the absorption coefficient (α). Its value is calculated from the following equation:

$$F(R) = \alpha = \frac{(1 - R)^2}{2R}$$

where F(R) is the Kubelka-Munk function, (α) is the absorption coefficient, and R is the diffuse reflectance [33].

The direct and indirect optical band gaps were both determined from their corresponding $[F(R)hv]^2$ and $[F(R)hv]^{1/2}$ graphical representations. The direct band gap energy values were estimated by extrapolation for pure and Ni/Zn doped $MnFe_2O_4$ are 3.47 and 3.53/3.36 eV, respectively. The indirect band gap calculated for pure and Ni/Zn doped $MnFe_2O_4$ are 3.21 and 3.32/3.38 eV, respectively. This suggests that the optical band gap of pure and with Ni/Zn doped $MnFe_2O_4$ in semiconductors changes depending on the nature of the element in the impurity.

3.3.2. Electrical Characterization

3.3.2.1. Frequency and temperature dependence of the complex impedance of manganese ferrite samples doped with Ni or Zn metal ions

The frequency dependence of the real (Z') and imaginary (Z'') components of the complex impedance $Z = Z' - iZ''$ (where $i = \sqrt{-1}$) of the investigated samples, in the frequency range 100 Hz-2 MHz and at temperatures of 30-120 °C, was measured using an RLC meter (Agilent type E4980A). The measurements were performed with the experimental setup shown in fig. 2-14 (see Ch. 2).

We used three samples for the measurements: a sample of manganese ferrite ($MnFe_2O_4$) marked sample I; a sample of manganese ferrite doped with Ni ions (sample J); and a sample of manganese ferrite doped with Zn ions (sample K). Fig. 3-7 a-f) presents the experimental results obtained for the three samples.

As can be seen in fig. 3-7 a), c) and e), for each constant temperature T, the real component of the complex impedance decreases with increasing frequency for all samples. In the approximate temperature range 32-60 °C for samples I (fig. 3-7 a)), J (fig. 3-7 c)), and K (fig. 3-7 e)), the real component (Z') (at low frequencies) decreases with increasing temperature, after which it increases slightly at 70-80 °C. It then decreases in the range 80-128 °C for samples I and J, but increases for sample K. The imaginary component Z'' displays a maximum at frequency f_{max} for all samples and all 10 temperatures investigated, as shown in fig 3-7 b), d), and f). This behavior represents the existence of an electrical relaxation process in the sample [35], which is due to the presence of charge carriers in the investigated samples [36, 37]. The maximum position of the component Z'' allows us to determine the relaxation time (τ) using the Debye relation, $2\pi f_{max}$ [34], for each temperature investigated. The values obtained for the relaxation times corresponding to the samples at the investigated temperatures are given in table 3-2.

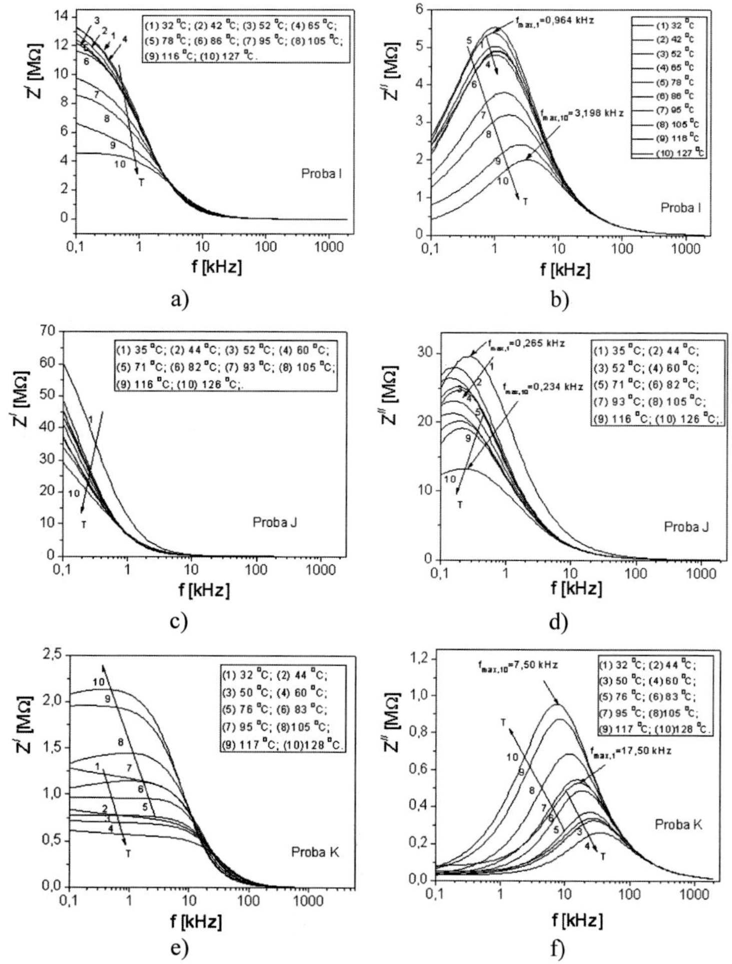

Fig. 3-7. Frequency and temperature dependence of the components of the complex impedance of the samples of undoped $MnFe_2O_4$ (sample I) and $MnFe_2O_4$ doped with Ni (sample J) or Zn (sample K).

Table 3-2. Relaxation time values by temperature.

Sample I		Sample J		Sample K	
$T\ [^0C]$	$\tau[\mu s]$	$T\ [^0C]$	$\tau[\mu s]$	$T\ [^0C]$	$\tau[\mu s]$
32	165,18	35	600,89	32	9,10
42	156,91	44	995,22	44	5,99
52	149,58	52	860,73	50	5,38
65	159,44	60	776,76	60	4,48
78	144,40	71	1120	76	6,43
86	151,22	83	970,95	83	8,84
95	113,00	93	1030	95	10,74
105	90,85	105	788,29	105	13,70
116	57,93	116	710,87	117	18,40
127	46,74	126	680,49	128	21,23

As can be seen in table 3-2, for sample I (manganese ferrite $MnFe_2O_4$), the relaxation time decreases with temperature both at the beginning of the range between 32 and 65 °C, and at high temperatures between 86 and 127 °C. In the case of sample J (manganese ferrite doped with Ni ions), the decrease in the relaxation time occurs at high temperatures above 70 °C. In the case of sample K (manganese ferrite doped with Zn ions), a decrease in the relaxation time takes place at the beginning of the temperature range 32-70 °C. This decrease in time for the mentioned temperature ranges accords with the Arrhenius law [36] (see eq. (2.19), Ch. 2). As shown in Chapter 2, the activation energy of the samples is due to the relaxation processes: $E_{A,relax}$ can be determined based on a straight line fit of the experimental dependence $ln\tau(T^{-1})$ (eq. (2.20), Ch. 2) for each investigated sample, as shown in fig. 3-8. a), b), and c).

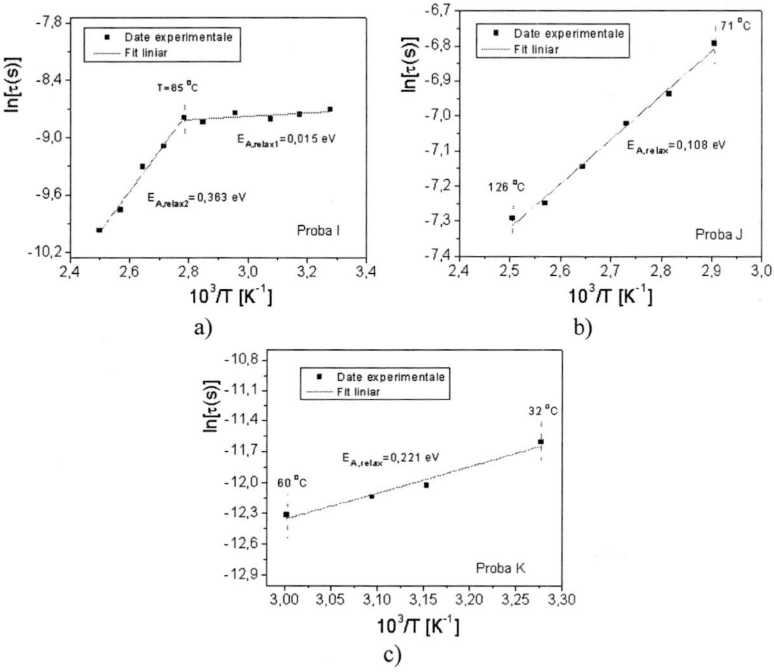

Fig. 3-8. The dependence of *lnτ* on T^{-1} for: (a) sample I of ferrite $MnFe_2O_4$; (b) sample J of $MnFe_2O_4$ ferrite doped with Ni ions; and (c) sample K of $MnFe_2O_4$ ferrite doped with Zn ions.

The values obtained for the three samples corresponding to the temperature ranges mentioned above are: $E_{A,relax1} = 0.015\ eV$ and $E_{A,relax2} = 0.363\ eV$ for sample I; $E_{A,relax} = 0.108\ eV$ for sample J; and $E_{A,relax} = 0.221\ eV$ for sample K. This is consistent with the values obtained in [39] by other authors for samples of the same type ($MnFe_2O_4$).

3.3.2.2. Complex electrical conductivity

The electrical conduction processes caused by the transport of charge carriers (electrons/voids or cations/anions) in $MnFe_2O_4$ samples doped with metal ions (*Me = Ni, Zn*) were analyzed in terms of their electrical conductivity. Under the action of the applied electric field, the charge carriers can migrate, thus contributing to the electrical response of the system [41]. The occurrence and behavior of AC conductivity has been explained by Jonscher [42], using the following relation:

$$\sigma = \sigma_{DC}(T) + \sigma_{AC}(\omega, T) \qquad (3.1)$$

where σ is the total conductivity; $\sigma_{DC}(T)$ is the continuous component (*DC-component*) of electrical conductivity, independent of frequency, but dependent on temperature; and $\sigma_{AC}(\omega, T)$ is the AC conductivity, depending on both frequency and temperature [38, 43].

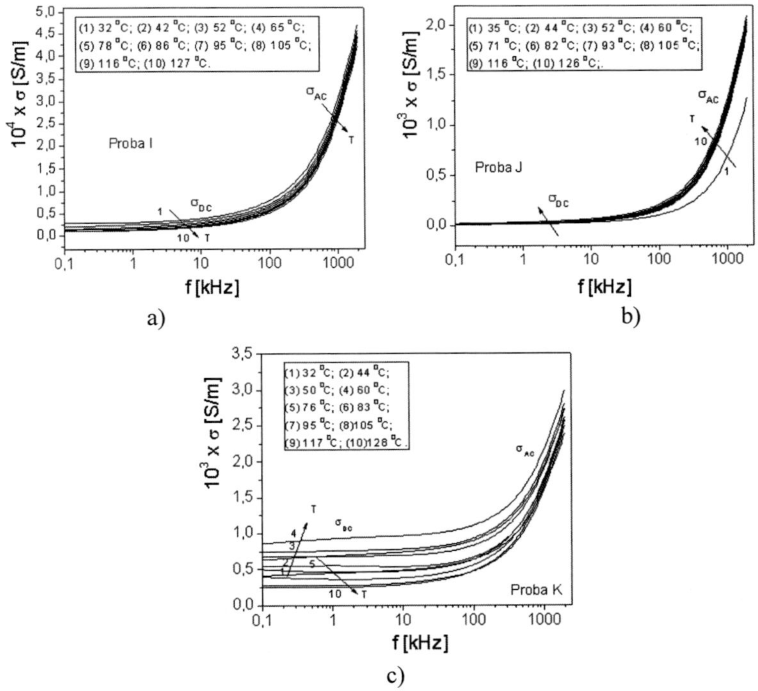

Fig 3-9. Frequency dependence of total conductivity σ at different temperatures for: a) sample I ($MnFe_2O_4$); b) sample J ($MnFe_2O_4$ doped with Ni ions); c) K sample ($MnFe_2O_4$ doped with Zn ions).

Using the obtained experimental values of (Z') and (Z'') for all three samples from fig. 3.7 and eq. (2.13) (see Ch. 2), we determined the total conductivity σ, which is the same as the real component σ' of the samples for each temperature T. Figure 3-9 shows the frequency dependence of the total conductivity σ for the three samples. The conductivity spectrum in fig. 3-9 can be divided into two regions that are in agreement with

Jonscher's universal law [42]: i) a linear region at low frequencies, which shows frequency-independent behavior (the first component of static or direct current conductivity σ_{DC}); and ii) a region of dispersion corresponding to a sudden increase in conductivity at high frequencies (the second component of conductivity σ_{AC} in alternating current). As the frequency of the field increases, the ferrite nanoparticles become more active in terms of electrical conduction due to the jumping of electrons between Fe^{2+} and Fe^{3+} ions, or the gaps between Ni^{2+} and Ni^{3+} in the octahedral interstices (B) [39, 46]. As a result, the conductivity σ_{AC} increases with increasing frequency.

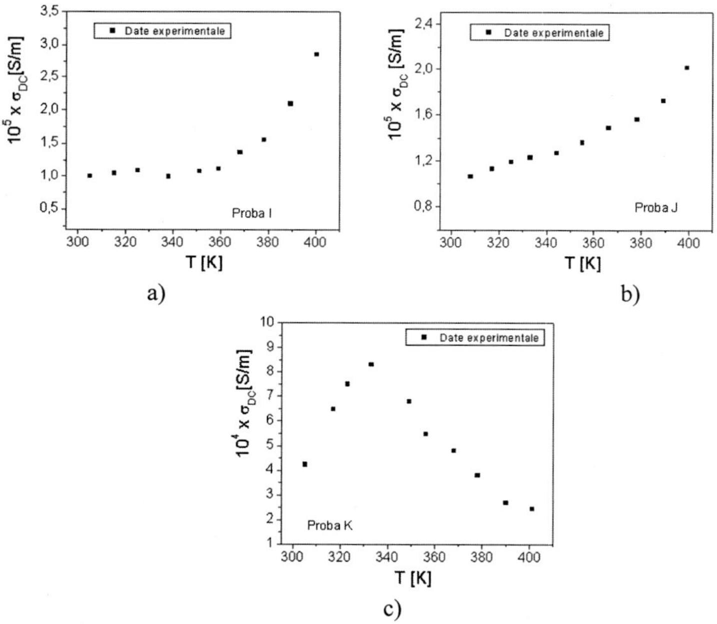

Fig. 3-10. Temperature dependence of static conductivity: a) sample I ($MnFe_2O_4$); b) test J ($MnFe_2O_4$ doped with Ni ions); and c) test K ($MnFe_2O_4$ doped with Zn ions).

Based on the values obtained for the static conductivity σ_{DC}, as shown in fig. 3-9, at all temperatures *T* at which the measurements were performed, we graphically represented the temperature dependence of the static conductivity $\sigma_{DC}(T)$, shown in fig. 3-10 a), b), and c) for the three investigated samples. As can be seen in fig. 3-10, the static conductivity

σ_{DC} values of manganese ferrite doped with Ni ions (sample J) are of the same order of magnitude as the σ_{DC} values of non-doped manganese ferrite $MnFe_2O_4$ (sample I). In the case of Zn ion doping of manganese ferrite (sample K), the conductivity σ_{DC} of the ferrite increases by an order of magnitude compared to the static conductivity σ_{DC} of the undoped manganese ferrite.

At the same time, in fig. 3-10 a) and b), we can see that for samples I ($MnFe_2O_4$) and J ($MnFe_2O_4$ doped with Ni ions), the component $\sigma_{DC}(T)$ increases slightly when the temperature increases from 30-70 °C, after which the increase is more pronounced at temperatures of 70-127 °C, indicating that the conduction process is thermally activated in these two temperature ranges. At the same time, the increase in conductivity in direct current σ_{DC} with temperature is due to an increase in the drift mobility of the thermally activated charge carriers and in accordance with the hopping conduction mechanism [47]. In the case of sample K ($MnFe_2O_4$ doped with Zn ions), the conductivity of $\sigma_{DC}(T)$ increases when the temperature increases from 30 °C to about 60 °C (fig. 3-10 c)), before decreasing as the temperature increases from 60 to 128 °C. The behavior of the K sample is in agreement with that shown by Mott and Davis [47]. Thus, in the low temperature range, the electrons are not free and the conductivity increases with increasing temperature. However, at temperatures above 60 °C, the electrons all become free and, as a result, the static conductivity decreases with increasing temperature.

3.4. Conclusion

Spinel ferrites of MFe_2O_4, undoped and doped with Ni/Zn, were prepared by the hydrothermal method. X-ray diffraction of the ferrite powders confirmed the formation of a spinel-phase crystalline structure. We observed that the lattice parameter differed when doped with Ni or Zn compared to the undoped sample. The FTIR spectra exhibited characteristic absorption bands for materials with a spinel structure; SEM images showed spherical particles of small size and agglomerates; while the EDX spectra showed the presence of dopants in the samples. The direct band gap energy was calculated for pure and doped $MnFe_2O_4$ (3.47 and 3.53/3.36 eV, respectively) from diffuse reflectance spectra using the Kubelka-Munk equation.

The σ_{DC} component for both undoped and doped ferrite samples remained approximately constant up to 10 kHz at each of the investigated temperatures. The results show that the value of the static conductivity σ_{DC} for manganese ferrite doped with Zn ions (sample K) is an order of magnitude higher than the conductivity σ_{DC} of the other samples. The σ_{DC} component for both undoped and doped ferrite samples remained approximately constant up to 10 kHz at each of the investigated temperatures. The results show that the value of the static conductivity σ_{DC} for manganese ferrite doped with Zn ions (sample K) is an order of magnitude higher than the conductivity σ_{DC} of the other samples. For $MnFe_2O_4$ doped with Ni ions, the $\sigma_{DC}(T)$ component increased slightly as the temperature rose from 30 °C up to about 70 °C. This increase is more pronounced at temperatures above 70 °C, indicating that the conduction process is thermally activated in the two temperature ranges, with the process taking place according to the hopping conduction mechanism.

References

[1] M. S. Park, J. Kim, K. J. Kim, J. W. Lee, J. H. Kim, Y. Yamauchi. Porous nanoarchitectures of spinel-type transition metal oxides for electrochemical energy storage systems. *Phys. Chem.* 17 (2015), 30963–30977.

[2] J. Smit and H. P. J. Wijn. Ferrites: physical properties of ferrimagnetic oxides in relation to their technical applications. Philips Technical Library, Eindhoven, Netherland, (1959) 149, 150.

[3] A. K. Singh, T. C. Goel, R. G. Mendiratta, O. P. Thakur and C. Prakash. Magnetic properties of Mn-substituted Ni-Zn ferrites. *J. Appl. Phys.* 92 (2002) 3872–3876.

[4] D. Makovec, A. Kodre, I. Arcon, M. Drofenik. Structure of manganese zinc ferrite spinel nanoparticles prepared with co-precipitation in reversed microemulsions. *J. Nanopart.* 11 (2009) 1145–1158.

[5] T. Giannakopoulou, L. Kompotiatis, A. Kontogeorgakos and G. Kordas. Microwave behavior of ferrites prepared via the sol-gel method. *J. Magn. Magn. Mater* 246 (2002) 360–365.

[6] J. V. Bellini, S. N. de Medeiros, A. L. L. Ponzoni, F. R. Longen, M. A. C. de Melo, and A. Paesano. Manganese ferrite synthesized from Mn(II) acetate + hematite freeze-dried powders. *Mater. Chem. Phys.* 105 (2007) 92–98.

[7] A. Manikandan, R. Sridhar, S. A. Antony, C. Meganathan, S. Ramakrishna, M. Bououdina, G. Sekaran. Effects of Mn^{2+} doping on structural, morphological, optical and catalytic properties of magnetic $CoFe_2O_4$ nanoparticles prepared by urea-assisted auto combustion method. *Ceram. Int.* 5 (2014) 122–133.

[8] S. Irfan, M.A.U. Nabi, Y. Jamil, N. Amin. Synthesis of Mn1-x Znx Fe_2O_4 ferrite powder by co-precipitation method. *IOP Conf. Series: Materials Science and Engineering* 60 (2014) 012048. doi:10.1088/1757-899X/60/1/012048.

[9] O. Karaagac, H. Köçkar. The effects of temperature and reaction time on the formation of manganese ferrite nanoparticles synthesized by hydrothermal method. *J Mater Sci: Mater. Electron* 31 (2020) 2567–2574.

[10] H. Anwar, A. Maqsood. Comparison of structural and electrical properties of Co2+ doped Mn-Zn soft nano ferrites prepared via coprecipitation and hydrothermal methods. *Mater. Res. Bull.* 49 (2014) 426–433.

[11] R. Topkaya, U. Kurtan, A. Baykal, M. S. Toprak. Polyvinylpyrrolidone (PVP)/$MnFe_2O_4$ nanocomposite: sol-gel autocombustion synthesis and its magnetic characterization. *Ceram. Int.* 39 (2013) 5651–5658.

[12] M. Stoia, C. Muntean, B. Militaru. Fine $MnFe_2O_4$ nanoparticles for potential environmental applications. *J. Therm. Anal. Calorim.* 121 (2015) 1003–1010.

[13] I. Desai, M. N. Nadagouda, M. Elovitz, M. Mills, B. Boulanger. Synthesis and characterization of magnetic manganese ferrites. *Materials Science for Energy Technol.* 2 (2019) 150-160.

[14] A. Das, K. Natarajan, S. Tiwari, A. K. Ganguli. Nanostructures synthesized by the reverse microemulsion method and their magnetic properties. *Mater. Res. Express* 7 (2020) 104001.

[15] S. Jadhav, B. Kim, H. Lee, I. Im, A. Rokade, S. Park, B.-S. Chun, G. Kim, Y.S. Yu, S. Lee. Induction heating and in vitro cytotoxicity studies of MnZnFe2O4 nanoparticles for self-controlled magnetic particle hyperthermia. *J. Alloy. Compd.* 745 (2018) 282–291.
[16] J. Mohapatra, A. Mitra, H. Tyagi, D. Bahadur, M. Aslam. Iron oxide nanorods as high-performance magnetic resonance imaging contrast agents. *Nanoscale* 7 (2015) 9174–9184.
[17] B. Zhou, Q. Wu, M. Wang, A. Hoover, X. Wang, F. Zhou, R.A. Towner, N. Smith et al. Immunologically modified $MnFe_2O_4$ nanoparticles to synergize photothermal therapy and immunotherapy for cancer treatment. *Chem. Ing. J.* 396 (2020) 125239.
[18] K. M. Katubi, Norah Salem Alsaiari, F. Alzahrani, S. M. Siddeeg, M. Tahoon. Synthesis of Manganese Ferrite/Graphene Oxide Magnetic Nanocomposite for Pollutants Removal from Water. *Processes* 9 (2021) 589. https://doi.org/10.3390/pr9040589.
[19] J. Chen, W. Wen, L. Kong, S. Tian, F. Ding, Y.Xiong. Magnetically Separable and Durable $MnFe_2O_4$ for Efficient Catalytic Ozonation of Organic Pollutants. *Ind. Eng. Chem. Res.* 53 (2014) 6297–6306.
[20] E. Skliri, I. Vamvasakis, I. T. Papadas, S. A. Choulis, G. S. Armatas. Mesoporous Composite Networks of Linked MnFe2O4 and ZnFe2O4 Nanoparticles as Efficient Photocatalysts for the Reduction of Cr(VI). *Catalysts* 11 (2021) 199.
[21] P. Yin, L. Zhang, P. Sun, J. Wang, X. Feng, Y. Zhang, J. Dai, Y. Tang. Apium-derived biochar loaded with MnFe2O4@ C for excellent low frequency electromagnetic wave absorption. *Ceram. Int.* 46 (2020) 13641–13650.
[22] I. Ismail, K. A. Matori, Z. Abbas et al. Single and Double-Layer Microwave Absorbers of Cobalt Ferrite and Graphite Composite at Gigahertz Frequency. *J. Supercond. Nov. Magn.* 32 (2019) 935–943.
[23] S. Sharifi, K. Rahimi, A. Yazdani. Highly improved supercapacitance properties of MnFe2O4 nanoparticles by MoS2 nanosheets. *Scientific Reports* 11 (2021) 8378. https://doi.org/10.1038/s41598-021-87823-6.
[24] M. N. Akhtar, A. A. Khan, M. N. Akhtar, M. Ahmad, M. A. Khan. Structural rietveld refinement, morphological and magnetic features of Cu doped Co-Ce nanocrystalline ferrites for high frequency applications. *Physica B: Condensed Matter.* 561 (2019) 121–131.
[25] U. Holzwarth, N. Gibson. The Scherrer equation versus the Debye-Scherrer equation. *Nature Nanotech* 6 (2011) 534. https://doi.org/10.1038/nnano.2011.145.
[26] X.-Y. Long, J.-Y. Li, D, Sheng, H.-Z. Lian. Spinel-type manganese ferrite ($MnFe_2O_4$) microspheres: A novel affinity probe for selective and fast enrichment of phosphopeptides. *Talanta* 166 (2017) 36–45.
[27] M. Kalyan Raju. FT-IR Studies of Cu Substituted Ni-Zn Ferrites for Structural and Vibrational Investigations. *Chem. Sci. Trans.* 4 (2015) 137–142.
[28] R. P. Patil, S. D. Delekar, D. R. Manec, P. P. Hankare. Synthesis, structural and magnetic properties of different metal ion substituted nanocrystalline zinc ferrite. *Results Phys.* 3 (2013) 129–133.

[29] M. Penchal Reddy, X. B. Zhou, A. Yann, D. Shiyu, Q. Huang. Low temperature hydrothermal synthesis, structural investigation and functional properties of $Co_xMn_{1-x}Fe_2O_4$ (0x1.0) nanoferrites. *Superlattices and Microstruct.* 81 (2015) 233–242.

[30] A. Mary Jacintha, A. Manikandan, K. Chinnaraj, S. A. Antony and P. Neeraja. Comparative Studies of Spinel $MnFe_2O_4$ Nanostructures: Structural, Morphological, Optical, Magnetic and Catalytic Properties. *J. Nanosci. Nanotechnol.* 15 (2015) 9732–9740.

[31] P. Hu, H. Yang, D. Pan, H. Wang, J. J. Tian, S. Zhang, X. Wang, A. A. Volinsky. Heat treatment effects on microstructure and magnetic properties of Mn–Zn ferrite powders. *Journal of Magnetism and Magnetic Materials* 322 (2010) 173–177.

[32] E. Girgis, M. M. S. Wahsh, A. G. M. Othman, L. Bandhu and K. V. Rao. Synthesis, magnetic and optical properties of core/shell $Co_{1-x}Zn_xFe_2O_4/SiO_2$ nanoparticles. *Nanoscale Res. Lett.* 6 (2011) 1–8.

[33] A. Mary Jacintha, A. Manikandan, K. Chinnaraj, S. Arul Antony, P. Neeraja. Comparative Studies of Spinel $MnFe_2O_4$ Nanostructures: Structural, Morphological, Optical, Magnetic and Catalytic Properties. *J. Nanosci. Nanotechnol.* 15 (2015) 9732–9740.

[34] P. Debye. *Polar Molecules*. PhD, p. 172. New York: Chemical Catalog Co., Inc. (1929).

[35] M. Nadeem, M. J. Akhtar, A. Y. Khan. *Solid State Commun.* 134 (2005) 431.

[36] A. K. Jonscher. *Nature* 267 (1977) 673.

[37] C.K. Suman, K. Prasad, R. N. P. Choudhary. *J. Mater. Sci.* 41(2006) 369–375

[38] P. Sfirloaga, I. Malaescu, M. Poienar, M. C. Nicolae, D. Malaescu, P. Vlazan. Synthesis, structural and electrical properties of $NaTaO_3$:Cu, *J. Mater. Sci: Mater. Electron.* doi:10.1007/s10854-016-5299-2.

[39] Dalibor L. Sekulić, Zorica Z. Lazarević, Čedomir D. Jovalekić, Aleksandra N. Milutinović, Nebojša Z. Romčević. Impedance Spectroscopy of Nanocrystalline $MgFe_2O_4$ and $MnFe_2O_4$ Ferrite Ceramics: Effect of Grain Boundaries on the Electrical Properties. *Science of Sintering*, 48 (2016) 17–28

[40] Evgenij Barsoukov, J. Ross Macdonald (eds.). *Impedance Spectroscopy Theory, Experiment, and Applications*. John Wiley & Sons, Hoboken, New Jersey (2005).

[41] H. Rahmouni, M. Nouiri, R. Jemai, N. Kallel, F. Rzigua, A. Selmi, K. Khirouni, S. Alaya. *J. Magn. Magn. Mater.* 316 (2007) 23–28.

[42] A. K. Jonscher. *Universal Relaxation Law*. (1st ed.) Chelsea Dielectrics Press, London (1996) 198–200.

[43] M. A. el Hiti. *Journal of Magnetism and Magnetic Materials* 164 (1996) 187.

[44] K. Funke. *Prog. Solid State Chem.* 22 (1993) 111–195.

[45] C. G. Koops. *Phys. Rev.* 83 (1951) 121-124.

[46] A. S. Fawzi, A. D. Sheikh, V. L. Mathe. *J. Alloys Compd.* 502 (2010) 231–237.

[47] N. F. Mott and E. A. Davis. *Electronic Process in Non-Crystalline Materials*. Clarendon Press, Oxford (1979).

CHAPTER 4

CERAMIC MATERIALS WITH PEROVSKITE STRUCTURES

PAULA SFIRLOAGĂ, PAULINA VLAZAN,
CATALIN N. MARIN, IOSIF MALAESCU

This chapter focuses on perovskite materials with the formula ABO_3, with a rare earth or alkaline earth cation coordinated by 12 oxygen ions present at the A-site and a transition metal cation in an octahedral structure at the B site. These compounds constitute promising candidates for a variety of applications, including gas sensors, catalysts, permanent magnets, pigments, etc. due to their adaptable composition, excellent redox properties, and superior thermal stability. The physicochemical properties of perovskite materials are known for their flexible characteristics associated with their A and/or B-site substitution/doping capabilities, and for their stoichiometrically available structures. Significant improvement in the structural, electrical, optical, and chemical properties can be achieved in perovskite materials by substitution or doping with different cations, as described in this chapter.

4.1. Introduction

Perovskite-type materials constitute a wide family of inorganic compounds with related crystallographic structures and diverse chemical compositions based on the general formula ABO_3 [1]. These are a new class of materials that are considered to be the best alternatives for noble metals in many fields of research and industry due to their attractive features, availability, and comparable cost [2]. These oxides have been studied as a low-cost material substitute for noble metals in electrocatalysis [3]; for hydrocarbons and VOCs [4] in catalytic oxidation; and in SCR for NOx removal [5]. Perovskite oxides have a well-defined crystalline structure that can generally be represented as ABO_3. Cation A has a large ionic radius—generally a rare earth element, an alkaline earth metal, or an alkali metal—while cation B has a smaller ionic radius and is generally a transition metal from the 3d, 4d, or 5d group [6]. Perovskite-type oxides not only exhibit many excellent physical and chemical characteristics, but are also widely applied in supercapacitors [7], batteries [8], gas and humidity sensors [9], and electrocatalysts [10] etc. The superior properties of perovskite structures, i.e. thermal stability, ionic conductivity, electronic and magnetic properties, as well as redox behaviour, can be tuned by varying the A or B cations in the lattice; their relative concentrations enable wide applicability in electrochemical devices, such as metal-air batteries and fuel cells [11-14].

Due to their good chemical and physical properties, perovskite-type oxides have many useful characteristics including good catalytic activity towards the oxygen evolution reaction (OER) and the oxygen reduction reaction (ORR) [15]; electronic conductivity [16]; oxygen ion mobility over the crystal lattice [17]; super-magnetic characteristics [18]; and dielectric, metallic, and semiconducting behaviors [19]. Among the perovskites with an ABO_3 structure, lanthanum manganite ($LaMnO_3$) is well-known in solid oxide fuel cell applications. Lanthanum manganite based compounds are common materials for cathodes in solid oxide fuel cells because of their compatibility with yttria-stabilized zirconia electrolytes [20]. Lanthanum manganite ($LaMnO_3$) with perovskite structure are attracting interest as a material for cathodes in solid oxide fuel cells [21], giant magnetoresistance [22], and as catalysts [23] due to their unique electrical, magnetic, and catalytic properties.

The structure and properties of $LaMnO_3$ perovskites are strongly influenced by the method of synthesis used. Different synthesis techniques have been developed in order to get pure crystalline $LaMnO_3$ powders. These compounds have been synthesized by many methods, including the sol-gel method [24]; the ultrasonic method with immersed sonotrode [25]; the solid-state reaction method [26]; and solution combustion synthesis [27]. However, using the conventional solid-state reaction for preparing $LaMnO_3$ powders requires a high calcining temperature, which induces sintering and particle aggregation. Conversely, using the co-precipitation technique it is difficult to control all cations to achieve precipitation in the solution at the same time, resulting in composition segregation. Doping with new components is regarded as an effective method for the formation of high quality and stable inorganic perovskite. According to existing research, component engineering by doping with new ions could modify the surface of the perovskite crystal, or see incorporation into the crystal lattice, replacing one of its substituents [28-30].

Perovskite materials have promising applications in electrochemistry [31-33] and, specifically, in renewable energy sources such as fuel cells and metal-air batteries. The current performance of these devices is strongly limited by the sluggish electrochemical reactions occurring suggesting the presence of molecular oxygen, i.e. the activity of the oxygen reduction reaction (ORR) and the oxygen evolution reaction (OER). Classical electrocatalysts to drive the ORR can be constituted by platinum-based materials, such as the alloys of platinum with Ag, Pd, and Ni to enhance their activity and/or stability. Alternatively, electrocatalysts can be made of platinum-based nanomaterials supported on a carbon base, which improves its surface area and increases the overall efficiency of the ORR [34, 35]. Perovskite oxides based on the ABO_3 crystal structure have been proposed in recent years as efficient bifunctional electrocatalysts for both the ORR and OER reactions. This is because they are low-cost materials the electrocatalytic properties of which can be tuned to exhibit better performance in relation to these reactions [36-38].

The oxygen reduction reaction (ORR), which is largely responsible for voltage losses in proton exchange fuel cells, is an enabling process for this attractive technology [39]. Transition metal oxides from the perovskite family have been proposed as catalysts for the ORR and OER, representing a low cost substitute for noble metals already by the 1970s. They are currently attracting increasing attention [40, 41]. The perovskite structure can accommodate various cations, resulting in variable surface and bulk properties [42]. Suntivich et al. have demonstrated that perovskites alone show low electrocatalytic activity [43], but this property may be increased by orders of magnitude if carbon is added to the catalytic layer.

4.2. Preparation of Perovskitic Materials by the Sol-gel Method and Physicochemical Characterization of the Obtained Materials

Undoped and doped (Ag, Pd, and N) $LaMnO_3$ materials were synthesized by the sol-gel method. To achieve the above mentioned materials, the following materials and reagents were used: lanthanum nitrate hexahydrate, $La(NO_3)_3 \cdot 6H_2O$; manganese nitrate tetrahydratenitrate, $Mn(NO_3)_2 \cdot 4H_2O$; silver nitrate, $AgNO_3$; palladium(II) nitrate dehydrate, $Pd(NO_3)_2 \cdot 2H_2O$; and urea (for N doping), sodium hydroxide (NaOH), and citric acid as a chelating agent.

To obtain undoped and doped $LaMnO_3$, stoichiometric amounts of $La(NO_3)_3 \cdot 6H_2O$ and $Mn(NO_3)_2 \cdot 4H_2O$ were dissolved in 50 mL bidistilled water. After adding the appropriate dopants, 1 g citric acid (chelating agent) and 50 mL alcohol were added.With continuous stirring, 2.5 mol/L NaOH was added until an alkaline medium (pH = 12) was reached, as required for perovskite phase precipitation.The obtained gel was dried in an oven at 80 °C for 10 h. The obtained powders were ground in an agate mortar and then annealed at 600 °C for 6 h with a heating rate of 5 °C/min.

Phase identification of the samples was performed by X-ray diffraction using a PANalyticalX'Pert Pro powder diffractometer with Cu-Ka radiations (l = 0.15406 nm) in a 2θ range from 10 to 80°, and a scanning rate of 2°/min. The average size of the crystallites was determined using the Scherrer equation. Rietveld refinements of the powder diffractograms were performed with the FullProf program (FullProf suite). The FT-IR spectra of the samples were recorded on a JASCO FT/IR-4200 apparatus using KBr pellets in the range 400-4000 cm^{-1}. The AFM analyses were recorded using the NanonicsMultiView 2000 scanner in intermittent mode and an ambient temperature of 24-25 °C. The morphology of the powders was investigated by SEM using an FEI Inspect-S scanning electron microscope, while elemental analysis and mapping was performed by energy dispersive X-ray analysis (EDX).

4.2.1. XRD analysis

Figure 4-1 shows the XRD patterns of undoped, silver, palladium, and nitrogen $LaMnO_3$ perovskite materials synthesized by the sol-gel method followed by thermal treatment at 600 °C for 6 h. The XRD patterns show that all the samples were well crystallized, as all the peaks were sharp and well defined. The crystalline structures and sizes of the powders were determined by Rietveld refinements using the FullProf program.

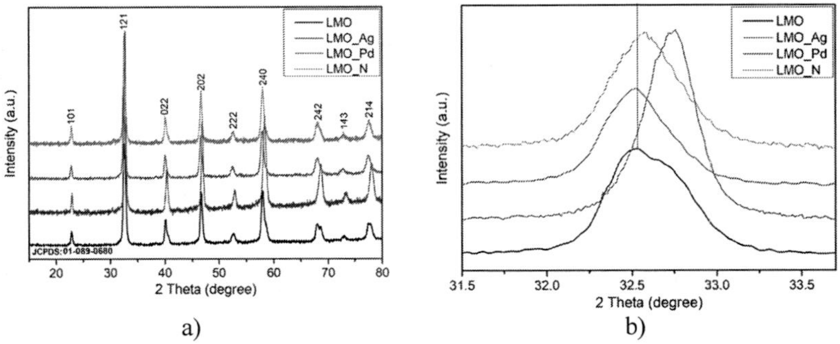

Fig. 4-1. XRD patterns for undoped $LaMnO_3$ and Ag, Pd, and N-doped $LaMnO_3$ (a). The highest intensity peak corresponds to 2θ = 32.5° (b).

The XRD diffraction peaks are well defined and correspond to the perovskite phase with an orthorhombic structure and a Pnma space group, indexed JCPDS: 01-089-0680, without any secondary phases. The crystallographic data determined by the refinement of the X-Ray data with the Rietveld method are presented in table 4-1. We can see that the size of the crystallites increases in the case of the three doped materials, indicating that the dopant elements entered the crystal lattice of the pure material. Moreover, the highest intensity peak, corresponding to 2θ = 32.5°, is shifted to the right, which confirms the successful doping of the $LaMnO_3$-type perovskite material.

Table 4-1. Crystallographic data for the studied materials.

Sample	Lattice parameters	Space group	Unit Cell Volume [Å^3]	Crystallite size [nm]
Undoped $LaMnO_3$	a=5.459(1) b=7.801(1) c=5.526(1)	P n m a	235.316	24.3
Ag doped $LaMnO_3$	a=5.455(1) b=7.728(1) c=5.5063(9)	P n m a	232.128	32.4
Pd doped $LaMno_3$	a=5.478(1) b=7.820(1) c=5.529(1)	P n m a	236.837	26
N doped $LaMnO_3$	a=5.4568(9) b=7.725(1) c=5.4988(7)	P n m a	231.790	35.2

4.2.2. FT-IR analysis

Figure 4-2 presents the FT-IR spectra for the studied sample obtained in the 400-400 cm^{-1} range. Several absorption bands were observed at wavelengths of 427, 619, 630, 1373, 1147, 1484, 1637, 2924, and 3437 cm^{-1}. Two transmission bands around 620 and 430 cm^{-1} were also observed. The band around 430 cm^{-1} corresponds directly to the bending mode (ν_b), owing to the change in the Mn–O–Mn bond angle. The frequency band around 620 cm^{-1} is related to the stretching mode (ν_s) of the Mn–O or Mn–O–Mn bonds, which embraces the change in the Mn–O bond's length due to internal motion [44, 45]. These two bands are related to the neighborhoods surrounding the $< MnO_6 >$ octahedral sites in the

ABO_3 perovskite and confirm the successful formation of a perovskite structure [46]. The FT-IR absorption band around 1490 cm^{-1} can be assigned to the asymmetric and symmetric stretching modes of C=O species adsorbed on the material's surface [47]. The bending and stretching mode of the OH bond of water was reflected in the FT-IR spectra around 1637 and 3437 cm^{-1}, respectively [48, 49].

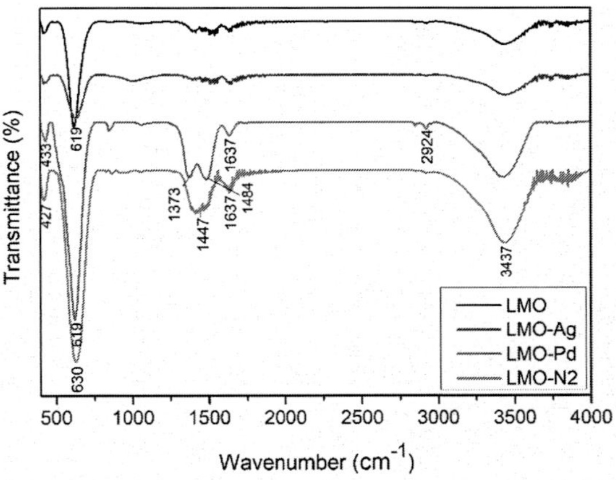

Fig. 4-2. FT-IR spectra for undoped and Ag, Pd, and N-doped $LaMnO_3$.

4.2.3. SEM and EDX results

Morphological evaluation of the studied perovskite oxide is shown in fig. 4-3. In order to confirm the successful doping of $LaMnO_3$, elemental composition of the samples was uncovered by EDX/Mapping analysis. From the surface morphological analysis, the presence of spherical particles can be observed that have a high tendency to agglomerate in aggregates of different sizes and shapes (especially in the case of undoped $LaMnO_3$ and Pd-doped $LaMnO_3$).

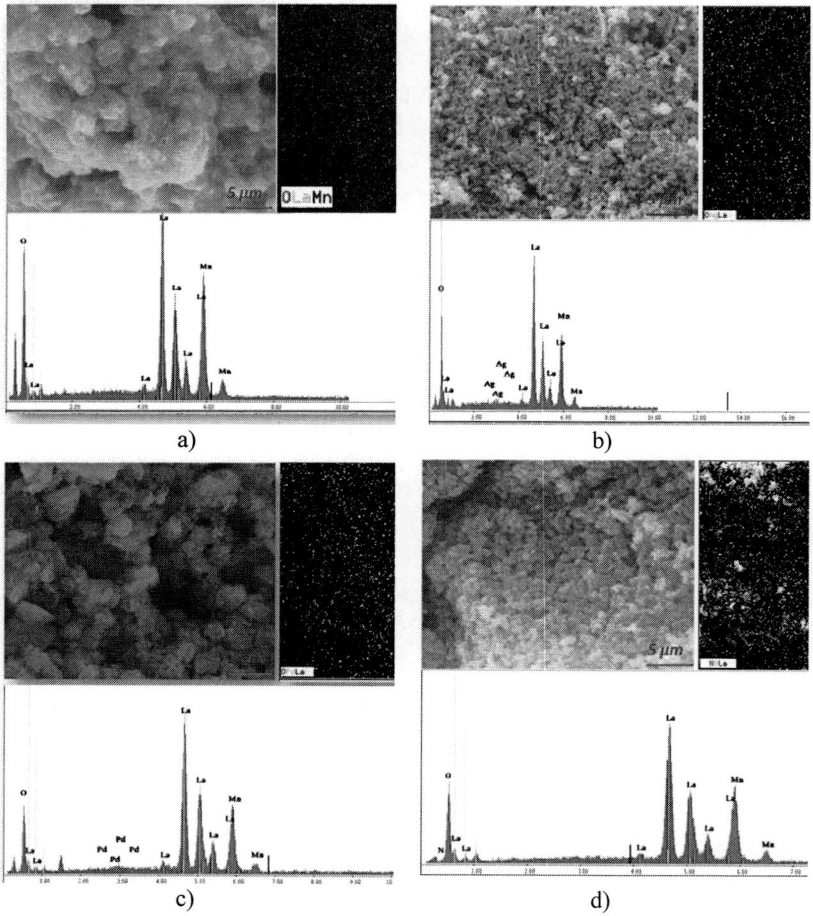

Fig. 4-3. SEM images of EDX and elemental mapping of: undoped $LaMnO_3$ (a); Ag-doped $LaMnO_3$ (b); Pd-doped $LaMnO_3$; and N-doped $LaMnO_3$.

Energy dispersive X-ray (EDX) confirmed the ratio of La, Mn, and O, as well as the phase purity of the obtained materials. The quantification of the elements shows that the stoichiometry of the perovskite compound had been preserved, but also that the dopant contents (Ag, Pd or N) were consistent with the amounts used in synthesis. These results indicate that the studied material (lanthanum manganite) was successfully doped using the sol-gel method. Moreover, elemental mapping was also carried out and the results show a homogeneous distribution of the elements La, Mn, and O, and of the dopants—Ag, Pd, or N–in the studied samples.

4.3. Electrocatalytic Activity of Electrodes Based on Perovskitic Compounds

Catalysts used for chemical and electrochemical reactions play an important role in modern technology and industry, with a range of applications from energy storage and conversion, to the reduction of toxic emissions and the synthesis of materials. In order to fulfill their role, catalysts must be very active, stable, and, at the same time, cheap. Such catalysts are also found in the class of compounds named perovskites [50].

Perovskites are mixed metal oxides of scientific and applied importance due to the physicochemical and electrochemical properties they possess, as well as being low in cost. These materials have been intensively studied especially in terms of their catalytic properties and have proven to be promising candidates for water electrolysis, oxygen reduction, and the oxidation of analytes present in aqueous solutions [51-53]. Moreover, the catalytic activity of perovskites can be enhanced by doping with metal ions, such as Ag and Pd [54-56].

Regarding water electrolysis, which is one of the most promising applications for non-polluting energy conversion [57], perovskites can be used both to catalyze the oxygen release reaction [58] and to catalyze the hydrogen release reaction [59].

Water electrolysis is a process that requires the decomposition of H_2O molecules into O_2 and H_2 [60], with two reactions taking place: the oxygen evolution of reaction, which takes place at the anode and leads to the decomposition of water to form O_2; and the reaction releases hydrogen, which takes place at the cathode and results in the formation of H_2 [61]. As can be seen for reactions (I) and (II), the hydrogen release reaction is easier to catalyze than the oxygen release reaction, because the evolution of oxygen requires more energy (in an alkaline environment, four electrons participate in the release of oxygen) [62-64].

$$2H^+ + 2e^- \rightarrow H_2 \quad \text{(I)}$$
$$4OH^- \rightarrow O_2 + 2H_2O + 4e^- \quad \text{(II)}$$

The theoretical potential of the oxygen release reaction is 1.23 V compared to the normal hydrogen electrode; however, in practice a higher value is required due to the kinetic barrier of the reactions [65-67].

The oxygen reduction reaction is considered a basic reaction for the conversion of electrochemical energy, playing a significant role in fuel cells, gas sensors, and the electrosynthesis of hydrogen peroxide. Oxygen reduction can take place either directly, with the involvement of four electrons (reaction III for alkaline solutions), or with the involvement of two electrons (reaction IV for alkaline solutions). Obtaining H_2O by oxygen reduction with the involvement of four electrons is used in both fuel cells and gas sensors, while H_2O_2 results from the oxygen reduction reaction with the involvement of two electrons.

$$O_2 + 2H_2O + 4e^- \rightarrow 4OH^- ; E^0 = 0.401 \text{ V} \quad \text{(III)}$$
$$O_2 + H_2O + 2e^- \rightarrow HO_2^- + OH^-; E^0 = -0.065 \text{ V} \quad \text{(IV)}$$

Reaction IV is followed by either another reduction reaction involving the participation of two electrons or by decomposition due to the disproportionate nature of the reaction. E° represents the standard potential of the electrode and the values are expressed with respect to the action of a normal hydrogen electrode at 25 °C [68].

The slow kinetics of the oxygen reduction reaction is a problem for fuel cells and metal-air batteries, as it leads to a decrease in the efficiency of these devices. Oxygen reduction can be catalyzed by perovskites or, more precisely, the transition metal oxides in the structure of perovskites are catalysts of the oxygen reduction reaction and can provide a catalytic activity comparable to that of rare metal catalysts, such as Pt [69, 70]. Among the perovskites that have been studied in this regard are undoped and Ag-doped $LaMnO_3$ [52].

Perovskites are also promising materials for the detection of analytes and are the subject of electrochemical detection studies reported in the literature, utilizing the property of perovskites to catalyze the oxidation of analytes [53, 71]. In such studies, powdered perovskitic materials were synthesized and their electrochemical properties investigated for the detection of anion nitrate (NO^{2-}). Successful detection is important because it reacts easily with proteins leading to the formation of carcinogenic N-nitrosamines, which cause esophageal cancer and methemoglobinemia. Although it is used as an antioxidant in food and beverages, the nitrite ion is toxic and its presence in drinking water should not exceed 65 μM. In addition, it leads to the eutrophication of closed water systems [53]. The amperometric detection of nitrite is based on its oxidation reaction with nitrate formation. An example of this is the deoxidation reaction of sodium nitrite leading to sodium nitrate (reaction V).

$$2NaNO_2 + O_2 \rightarrow 2NaNO_3 \quad \text{(V)}$$

The electrochemical studies described envisage the achievement of modified electrodes with undoped and doped perovskitic compounds having different concentrations of metal ions (Ag and Pd) in order to identify their electrocatalytic properties for oxygen release and reduction reactions, as well as for the oxidation reaction of nitrite added in different concentrations in the electrolyte solution. In the case of the oxygen release reaction, an optimization study was performed according to the preliminary results obtained. This involved the use of different conductive substrates (graphite and nickel), different dispersing agents (ethanol and dimethylformamide), and different binders (Nafion and PEDOT). The influence of conductive support, the type of dispersing agent, as well as the type and amount of binder on the electrochemical characteristics of the modified electrodes were all taken into account. The experimental results highlight the electrocatalytic performance of the doped perovskitic compounds, as well as the performance of hybrid materials (compositions containing other materials besides perovskites) used to modify the studied electrodes.

4.3.1. Experiments

4.3.1.1. Materials and reagents

Undoped and doped $LaMnO_3$ perovskites with different concentrations of Ag (0.05 %, 0.1 %, and 0.15 %) or Pd (0.05 %, 0.1 %, and 0.15 %) were synthesized by the sol-gel method.

Spectroscopic graphite (type SW.114, produced by “Kablo Bratislava”, National Corporation “Electrocarbon Topolcany” Factory, Slovakia) and high purity nickel plate were used as conductive substrates. Carbon black Vulcan XC 72 (Fuell Cell Store) was used to improve the transfer of electrons between the electroactive species and the electrode. For better fixation of the perovskites and the Vulcan carbon to the surface of the conductive substrates, two types of binders were used: Nafion® 117 solution of 5 % concentration in a mixture of lower aliphatic alcohols and water (Sigma-Aldrich) and a 0.8 % concentration PEDOT:PSS solution in water (Sigma-Aldrich). The N,N-dimethylformamide (DMF), potassium hydroxide, potassium chloride, potassium ferricyanide, potassium nitrate, sodium nitrite, ethanol, and acetone, all of analytical purity, were purchased from Sigma-Aldrich and Merk. Bidistilled water was used to obtain the solutions used in the studies.

4.3.1.2. Procedures for obtaining perovskite-modified electrodes

The drop-casting method deposition method was used to modify the graphite or Ni-conductive support and the suspensions that were deposited on the surface of the support contained either perovskitic material dispersed in ethanol or dimethylformamide (DMF), or a hybrid compound consisting of perovskitic material mixed with conductive carbon (Vulcan) and Nafion or PEDOT in the same dispersing agents. The modified electrodes were investigated for their electrocatalytic activity for the oxygen release reaction, for the oxygen reduction reaction, and for the oxidation reaction of an analyte (nitrite ion) present in the electrolyte solution.The procedures used to obtain the modified electrodes are described below.

- Procedures to modify electrodes with perovskites for catalysis studies in relation to the oxygen release reaction.

- **Procedures for the manufacture of electrodes modified with perovskites for catalysis studies in relation to the oxygen release reaction.**

- *Procedure 1*: Undoped or doped $LaMnO_3$ with different concentrations of Ag (0.05 %, 0.1 %, and 0.15 %) or Pd (0.05 %, 0.1 %, and 0.15 %) was added in the amount of 5 mg in 0.5 mL ethanol. The resulting suspension was subjected to ultrasonic action for 10 minutes, then transferred by drop-casting (10 μL) onto the surface of the graphite support and left to dry for 24 hours at room temperature.
- *Procedure 2*: Undoped or doped $LaMnO_3$ with different concentrations of Ag (0.05 %, 0.1 %, and 0.15 %) or Pd (0.05 %, 0.1 %, and 0.15 %) was added in the amount of 5 mg in 0.5 mL ethanol together with a volume of 10 μL Nafion solution. The resulting suspension was sonicated for 10 minutes, then transferred by drop-casting (10 μL) onto the surface of the graphite support and left to dry for 24 hours at room temperature.
- *Procedure 3*: Undoped or doped LaMnO3 with different concentrations of Ag (0.05 %, 0.1 %, and 0.15 %) or Pd (0.05 %, 0.1 %, and 0.15 %) was added in the amount of 5 mg in 0.5 mL ethanol, together with 2.5 mg Vulcan conductive carbon. The resulting suspension was sonicated for 10 minutes, then transferred by drop-casting (10 μL) onto the surface of the graphite support and left to dry for 24 hours at room temperature.

- *Procedure 4*: Undoped or doped LaMnO3 with different concentrations of Ag (0.05 %, 0.1 %, and 0.15 %) or Pd (0.05 %, 0.1 %, and 0.15 %) was added in the amount of 5 mg in 0.5 mL ethanol, together with 2.5 mg Vulcan conductive carbon and a volume of 10 μL Nafion solution. The resulting suspension was sonicated for 10 minutes, then transferred by drop-casting (10 μL) onto the surface of the graphite support and left to dry for 24 hours at room temperature.

Table 4-2 shows the compositions of the suspensions used, according to procedures 1-4, for the modification of the graphite conductive support in order to achieve perovskite-modified electrodes. Linear voltammetry experiments have shown that of the modified electrodes obtained by procedures 1-4, the best electrocatalytic properties for the oxygen release reaction were seen for the G/LMO_Ag3 electrode. This was further investigated by an optimization study aimed at improving electrocatalytic performance. The study considered: the influence of the conductive support using both the graphite and Ni support; the influence of the type of binder used—Nafion or PEDOT; as well as the influence of the amount of binder used (3 μl Nafion or PEDOT; 5 μL Nafion or PEDOT) on the electrocatalytic characteristics of the modified electrode. The compositions of the suspensions used in the manufacture of modified electrodes for optimization experiments, starting with the G/LMO_Ag3 electrode, are presented in table 4-3.

Quantities of 10 μL from the suspensions were applied to the surface of graphite and nickel substrates. This was followed by the solvent evaporation step, which lasted 24 hours. Evaporation took place at room temperature, except where the solvent used was DMF. To ensure evaporation of this solvent, a heat treatment of 50 °C was applied for 30 minutes, followed by drying at room temperature for 23 hours.

Table 4-2. Compositions of the suspensions used to modify the graphite conductive support in order to obtain electrodes modified with perovskite, according to procedures 1-4.

Procedure	*Name of the modified electrode*	*Type of perovskitic material*	*The amount of perovskitic material [mg]*	*The amount of conductive carbon Vulcan [mg]*	*The volume of Nafion solution [µL]*	*The volume of ethanol [µL]*
1	G/LMO	Undoped $LaMnO_3$	5	-	-	500
	G/LMO_Ag1	Ag 0,05% doped $LaMnO_3$	5	-	-	500
	G/LMO_Ag2	Ag 0,1% doped $LaMnO_3$	5	-	-	500
	G/LMO_Ag3	Ag 0,15% doped $LaMnO_3$	5	-	-	500
	G/LMO_Pd1	Pd 0,05% doped $LaMnO_3$	5	-	-	500
	G/LMO_Pd2	Pd 0,1% doped $LaMnO_3$	5	-	-	500
	G/LMO_Pd3	Pd 0,15% doped $LaMnO_3$	5	-	-	500

2	G/LMO_N	Undoped $LaMnO_3$	5	-	10	500
	G/LMO_Ag1_N	Ag 0,05% doped $LaMnO_3$	5	-	10	500
	G/LMO_Ag2_N	Ag 0,1% doped $LaMnO_3$	5	-	10	500
	G/LMO_Ag3_N	Ag 0,15% doped $LaMnO_3$	5	-	10	500
	G/LMO_Pd1_N	Pd 0,05% doped $LaMnO_3$	5	-	10	500
	G/LMO_Pd2_N	Pd 0,1% doped $LaMnO_3$	5	-	10	500
	G/LMO_Pd3_N	Pd 0,15% doped $LaMnO_3$	5	-	10	500
3	G/LMO_V	Undoped $LaMnO_3$	5	2,5	-	500
	G/LMO_Ag1_V	Ag 0,05% doped $LaMnO_3$	5	2,5	-	500
	G/LMO_Ag2_V	Ag 0,1% doped $LaMnO_3$	5	2,5	-	500
	G/LMO_Ag3_V	Ag 0,15% doped $LaMnO_3$	5	2,5	-	500
	G/LMO_Pd1_V	Pd 0,05% doped $LaMnO_3$	5	2,5	-	500
	G/LMO_Pd2_V	Pd 0,1% doped $LaMnO_3$	5	2,5	-	500
	G/LMO_Pd3_V	Pd 0,15% doped $LaMnO_3$	5	2,5	-	500
4	G/LMO_V_N	Undoped $LaMnO_3$	5	2,5	10	500
	G/LMO_Ag1_V_N	Ag 0,05% doped $LaMnO_3$	5	2,5	10	500
	G/LMO_Ag2_V_N	Ag 0,1% doped $LaMnO_3$	5	2,5	10	500
	G/LMO_Ag3_V_N	Ag 0,15% doped $LaMnO_3$	5	2,5	10	500
	G/LMO_Pd1_V_N	Pd 0,05% doped $LaMnO_3$	5	2,5	10	500
	G/LMO_Pd2_V_N	Pd 0,1% doped $LaMnO_3$	5	2,5	10	500
	G/LMO_Pd3_V_N	Pd 0,15% doped $LaMnO_3$	5	2,5	10	500

Table 4-3. Compositions of the suspensions used to modify the graphite or Ni conductive support in order to obtain electrodes modified with perovskite for the optimization study.

Conductive support	*Name of the modified electrode*	*Type of perovskitic material*	*The amount of perovskitic material [mg]*	*The volume of PEDOT [µL]*	*The volume of Nafion solution [µL]*	*The volume of ethanol [µL]*	*The volume of DMF [µL]*
Graphite	G/LMO_Ag3_DMF	Ag 0,15% doped $LaMnO_3$	5	-	-	-	500
	G/LMO_Ag3_N5		5	-	5	500	-
	G/LMO_Ag3_P5		5	5	-	500	-
	G/LMO_Ag3_DMF_N5		5	-	5	-	500
	G/LMO_Ag3_DMF_P5		5	5	-	-	500
	G/LMO_Ag3_N3		5	-	3	500	-
	G/LMO_Ag3_P3		5	3	-	500	-
	G/LMO_Ag3_DMF_N3		5	-	3	-	500
	G/LMO_Ag3_DMF_P3		5	3	-	-	500
Nickel	Ni/LMO_Ag3	Ag 0,15% doped $LaMnO_3$	5	-	-	500	-
	Ni/LMO_Ag3_DMF		5	-	-	-	500
	Ni/LMO_Ag3_N5		5	-	5	500	-
	Ni/LMO_Ag3_P5		5	5	-	500	-
	Ni/LMO_Ag3_DMF_N5		5	-	5	-	500
	Ni/LMO_Ag3_DMF_P5		5	5	-	-	500
	Ni/LMO_Ag3_N3		5	-	3	500	-
	Ni/LMO_Ag3_P3		5	3	-	500	-
	Ni/LMO_Ag3_DMF_N3		5	-	3	-	500
	Ni/LMO_Ag3_DMF_P3		5	3	-	-	500

- **Procedures used in the manufacture of electrodes modified with perovskites for studies of catalysis in relation to the oxygen reduction reaction.** The four procedures used to obtain modified electrodes with the desired electrocatalytic properties were studied in relation to the oxygen release reaction. They were also used in the investigation of the electrodes in relation to their electrocatalytic activity for the oxygen reduction reaction. As such, the suspensions were applied to the surface of the graphite substrate with the compositions presented in tables 1 and 28 to obtain modified electrodes.

- **Procedures used in the manufacture of electrodes modified with perovskites for catalysis studies in relation to the nitrous oxide reaction.**
- *Procedure 5*: Undoped $LaMnO_3$ (10 mg) was added to 0.5 mL DMF and the resulting suspension was subjected to ultrasound for 10 minutes, before being transferred by drop-casting (15 μL) onto the surface of the graphite support and left to dry for 48 hours at room temperature.
- *Procedure 6*: $LaMnO_3$ (undoped or doped with Ag or Pd in 0.15 % concentrations) and Vulcan conductive carbon were added in the amounts of 10 mg and 5 mg in 0.5 mL DMF, respectively. The resulting suspension was sonicated for 10 minutes, then transferred by drop-casting (15 μL) onto the surface of the graphite support and left to dry for 48 hours at room temperature.
- *Procedure 7*: Undoped $LaMnO_3$ (10 mg), Vulcan conductive carbon (5 mg), and 25 μL Nafion solution were added to 0.5 mL DMF. The resulting suspension was sonicated for 10 minutes, then transferred by drop-casting (15 μL) onto the surface of the graphite support and left to dry for 48 hours at room temperature.

Table 4-4 shows the compositions of the suspensions used according to procedures 5-7 for the modification of the graphite conductive support in order to make perovskite-modified electrodes.

Table 4-4. Compositions of the suspensions used to modify the graphite conductive support in order to obtain electrodes modified with perovskite according to procedures 5-7.

Procedure	*Name of the modified electrode*	*Type of perovskitic material*	*Type of perovskitic material [mg]*	*The amount of conductive carbon Vulcan [mg]*	*Nafion solution volume [μL]*	*DMF volume [μL]*
5	G/LMO_DMF	Undoped $LaMnO_3$	10	-	-	500
6	G/LMO_DMF_V	Undoped $LaMnO_3$	10	5	-	500
	G/LMO_Pd3_DMF_V	Pd 3 0,15% doped $LaMnO_3$	10	5	-	500
	G/LMO_Ag3_DMF_V	Ag3 0 15% doped $LaMnO_3$	10	5	-	500
7	G/LMO_DMF_V_N	Undoped $LaMnO_3$	10	5	25	500

4.3.1.3. Electrochemical experiments

A Voltalab PGZ 402 (Radiometer Analytical) potentiometer and a glass cell with three electrodes were used to perform the electrochemical experiments. The auxiliary electrode consisted of a Pt plate (Sgeom = 0.8 cm^2) and the Ag/AgCl electrode (sat. KCl) was used as a reference. Spectroscopic graphite or nickel, unmodified or modified with perovskites or compositions containing perovskites, was used as the working electrode. All experiments were performed at room temperature.

The spectroscopic graphite was in the form of a bar trapped in a polyethylene tube that was tightly bound to it by heat treatment (~180 °C). One of the two ends of the bar was used to connect the electrode to the potentiostat, while the other end was immersed in the electrolyte solution, ensuring a geometric surface of 0.28 cm^2. Before modification or use as in electrochemical experiments, the surface was prepared by grinding with abrasive papers of different granularities (600, 800, 1200), followed by polishing with felt, washing with water, ethanol, and acetone, and drying at room temperature. The conductive substrate of Ni was in the form of a disc cut from a nickel plate. It was sanded with abrasive paper of different granulations (600, 800, 1200) and polished with felt in the presence of an alumina suspension (particle size = 0.05 μm). It was then washed with detergent water, rinsed with double-distilled water, and dried by wiping with laboratory paper. The disc was held in a polyamide support connected to the potentiostat (Sgeom = 0.28 cm^2).

The current densities (i) referred to in the text are geometric current densities, unless otherwise specified. The potentials (E) obtained during the experiments are expressed either as a function of the Ag/AgCl electrode (sat. KCl), or as a function of the reversible hydrogen electrode (RHE). Equation (1) was used to express the potentials for RHE and the oxygen release superpotential (η) was calculated with equation (2) [72].

$$E_{RHE} = E_{Ag/AgCl(sat.KCl)} + 0{,}059 \times pH + 0{,}197 \qquad (4.1)$$

$$\eta = E_{RHE} - 1{,}23 \qquad (4.2)$$

where E_{RHE} is the potential of the reversible hydrogen electrode [V]; $E_{Ag/AgCl(sat.\ KCl)}$ is the potential expressed as a function of the reference electrode Ag/AgCl (sat. KCl) [V]; and η is the superpotential [V].

The experiments in which the electrocatalytic properties of the modified electrodes were studied in relation to the oxygen release reaction involved the recording of linear voltammograms with compensation of the voltage drop in the electrolyte solution. The electrolyte solution used was

0.1 M KOH (pH = 13) and the scanning speed (v) was 1 mV/s. The catalytic characteristics of the electrode chosen after the optimization study were also investigated using KOH with a concentration of 0.5 M (pH = 13.7), respectively 1 M (pH = 14). To estimate the electroactive surface of this electrode, cyclic voltammograms were recorded at different scanning speeds in a 1 M concentration KNO3 solution to which 4 mM K3 [Fe (CN) 6] was added. The experimental results were used in equation (3)—the Randles-Sevcik equation—by which the value of the diffusion coefficient of ferricyanide ions was also estimated [73].

$$I_p = (2{,}69 \times 10^5) \times n^{3/2} \times A \times D^{1/2} \times C \times v^{1/2} \quad (4.3)$$

where I_p is the peak current [A]; n is the the number of electrons involved in the redox process at $T = 298$ K (n = 1 for the ferrox/ferrocyanide redox couple); A is the working electrode surface area [cm^2]; D is the diffusion coefficient of the electroactive species (in the case of potassium ferricyanide solution, the theoretical value for D is 0.67 x 10^{-5} cm^2/s); C is the concentration of the electroactive species [M]; and v is the scanning speed [V/s].

The study of the electrocatalytic properties of the modified electrodes in relation to the oxygen reduction reaction involved the recording of cyclic voltammograms with compensation of the voltage drop in the electrolyte solution. Recordings were made to V = 5 mV/s in a 0.1 M KOH concentration. Before each experiment, O_2 was bubbled directly into the 20 mL electrolyte solution in which the electrodes were immersed for 10 minutes to facilitate tracing of the voltammograms.

Cyclic voltammetry was also used to investigate the electrocatalytic activity of the modified electrodes for the nitrous oxide reaction. The curves were recorded at v = 100 mV/s in a KCl solution with a concentration of 0.1 M. For each electrode, a volume of 5 mL electrolyte solution was started to which volumes of 10 or 40 μL of $NaNO_2$ solution were gradually added, taken from a 73 mM stock solution. Voltammograms were plotted in the absence of nitrite and after each volume of $NaNO_2$ solution was added.

4.3.2. Results and discussion

4.3.2.1. Investigation of the electrocatalytic activity of electrodes modified with perovskites for the oxygen release reaction

Figure 4-4 shows linear voltammograms using modified electrodes as described in procedure 1. At current densities greater than 14 mA/cm^2, we can observe that the curve obtained with the G/LMO_Ag1 electrode almost overlaps with the one recorded for the unmodified electrode (G), but does not reach a current density as high as G in the scanned potential range. The rest of the curves indicate that, for $i > 14$ mA/cm^2, the catalytic activity of the remaining modified electrodes is higher than that of the unmodified electrode. The only exception is observed for the G/LMO electrode, in which case the curve drawn overlaps with that obtained for the unmodified electrode at high values of current density. The voltammogram recorded for the electrode G/LMO_Ag3 stands out by the fact that it reaches a value of current density ($i = 43.8$ mA/cm^2) that is higher than the others.

Regarding the oxygen release overpotential (η), at a current density of 10 mA/cm^2, a value for which the overpotential [74, 75] is usually specified, it is 0.985 V; while at $i = 20$ mA/cm^2 the η value is 1.048 V. At these current densities, the curve obtained for the the G/LMO_Ag3 electrode overlaps with the curves using the G/LMO_Pd2 electrode (when $i = 10$ mA/cm^2) and the G/LMO_Pd1 electrode (when $i = 20$ mA/cm^2), whereas the mentioned η values are lowest for electrodes modified as described in procedure 1. Among the mandatory properties of the catalysts used to decompose water, current density must be as high as possible, accompanied by an overpotential value as low as possible [76, 77], and the G/LMO_Ag3 electrode was chosen as the most efficient in terms of catalytic view.

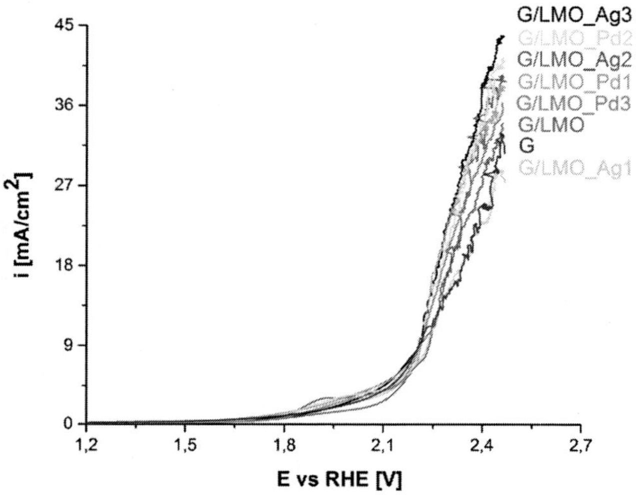

Fig. 4-4. Linear voltammograms recorded in a KOH solution with a concentration of 0.1 M at v = 1 mV/s, using the unmodified graphite electrode (G) and graphite electrodes modified by procedure 1: G/LMO, G/LMO_Ag1, G/LMO_Ag2, G/LMO_Ag3, G/LMO_Pd1, G/LMO_Pd2, and G/LMO_Pd3.

Figure 4-5 shows the recorded polarization curves of the electrodes modified by procedure 2. The results show that the addition of Nafion in the suspension employed for the modification of the electrodes decreases their electrocatalytic activity for the oxygen release reaction. Thus, we can observe that, in the considered potential range, the voltammogram of the unmodified electrode reaches a higher current density than for the curves obtained by using the modified electrodes. In addition, the release of oxygen at this electrode takes place at higher potentials for almost all current density values of the considered range.

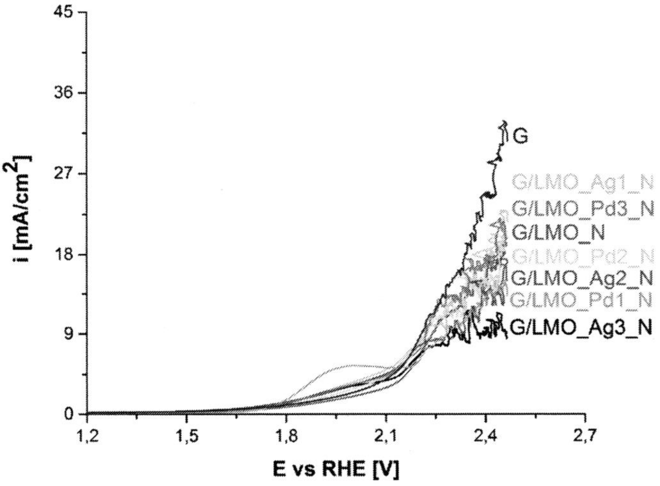

Fig. 4-5. Linear voltammograms recorded in a KOH solution with a concentration of 0.1 M at v = 1 mV/s, using the unmodified graphite electrode (G) and graphite electrodes modified by procedure 2: G/LMO_N, G/LMO_Ag1_N, G/LMO_Ag2_N, G/LMO_Ag3_N, G/LMO_Pd1_N, G/LMO_Pd2_N, and G/LMO_Pd3_N.

The recorded voltammograms of the electrodes modified by procedures 3 and 4 are shown in figs. 4-6 and 4-7. The electrocatalytic activity of the electrodes obtained using suspensions that contain Vulcan and Nafion, respectively, in addition to perovskites, is seen to be lower than that of the unmodified electrode. Comparing the polarization curves from fig. 1-4 of the modified electrodes with different compositions, but the same type of perovskitic material, a gradual decrease in electrocatalytic activity can be observed. As such, we can see that the electrodes modified by procedure 1 generally have the best electrocatalytic properties for the oxygen release reaction. The next best are those electrodes modified by procedure 2, followed by the ones modified by procedure 3. The weakest catalytic performance was seen in the case of electrodes modified by procedure 4. An example is given for electrodes modified by all four procedures with suspensions containing $LaMnO_3$ perovskite doped with 0.15 % Ag. The data extracted from the polarization curves and plotted for the respective electrodes are shown in table 4-5.

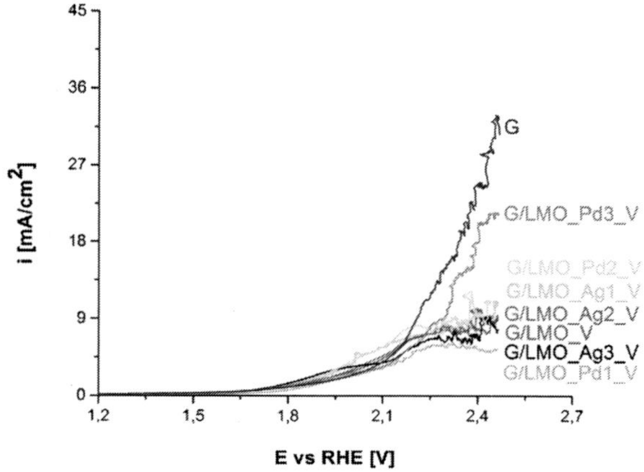

Fig. 4-6. Linear voltammograms recorded in a KOH solution with a concentration of 0.1 M at v = 1 mV/s, using the unmodified graphite electrode (G) and graphite electrodes modified by procedure 3: G/LMO_V, G/LMO_Ag1_V, G/LMO_Ag2_V, G/LMO_Ag3_V, G/LMO_Pd1_V, G/LMO_Pd2_V, and G/LMO_Pd3_V.

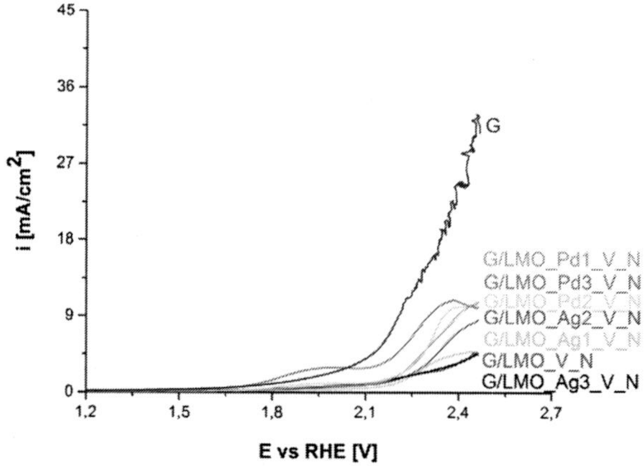

Fig. 4-7. Linear voltammograms recorded in a KOH solution with a concentration of 0.1 M at v = 1 mV/s, using the unmodified graphite electrode (G) and graphite electrodes modified by procedure 4: G/LMO_V_N, G/LMO_Ag1_V_N, G/LMO_Ag2_V_N, G/LMO_Ag3_V_N, G/LMO_Pd1_V_N, G/LMO_Pd2_V_N, and G/LMO_Pd3_V_N.

Table 4-5. Data from linear voltammograms recorded using electrodes modified with suspensions containing 0.15 % Ag-$LaMnO_3$ perovskite by the 4 procedures.

Name of the modified electrode	*Procedure for obtaining the modified electrode*	i_{max} *[mA/cm²]*	η *corresponding to* $i = 5$ *mA/cm² [V]*
G/LMO_Ag3	1	43,8	0,89
G/LMO_Ag3_N	2	11,65	0,93
G/LMO_Ag3_V	3	9,18	0,94
G/LMO_Ag3_V_N	4	4,48	-

In table 4-5, i_{max} represents the maximum current density reached in the potential range 0-1.5 V compared to Ag/AgCl (sat. KCl), for which the voltammograms in figs. 4-4 and 4-7 were recorded. There is a gradual decrease in values i_{max} and the gradual increase of the overpotential for i = 5 mA/cm^2, from the first to the last electrode modification procedure. The polarization curve obtained on the electrode modified by procedure 4 does not even reach i = 5 mA/cm^2 in the potential range considered. Modified graphite electrodes with electrocatalytic activity for the weaker oxygen release reaction than unmodified graphite, such as the G/LMO_Ag3_V_N electrode, can be used in electrochemical applications where oxygen release is desired with high superpowers. Thus, it is possible to investigate some gases on an unmodified graphite electrode that emit at superpowers higher than oxygen.

Of all the modified electrodes investigated, the G/LMO_Ag3 electrode has the best electrocatalytic properties for the oxygen release reaction and has been the subject of an optimization study. The modified electrodes made for this study were those presented in table 2 and the linear voltammograms recorded using these electrodes are shown in figs. 4-8 and 4-9. Thus, fig. 5 shows the polarization curves obtained on electrodes made by modifying the graphite substrate with $LaMnO_3$ perovskite doped with 0.15 % Ag applied (or in compositions with Nafion or PEDOT) from suspensions in which the dispersion medium was either ethanol or DMF.

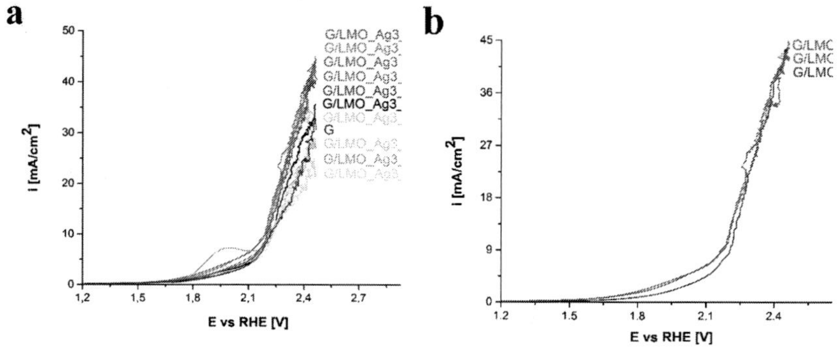

Fig. 4-8. (a) Linear voltammograms recorded in a KOH solution with a concentration of 0.1 M at v = 1 mV/s, using the unmodified graphite electrode (G) and the following modified graphite electrodes: G/LMO_Ag3, G/LMO_Ag3_DMF, G/LMO_Ag3_N5, G/LMO_Ag3_P5, G/LMO_Ag3_DMF_N5, G/LMO_Ag3_DMF_P5, G/LMO_Ag3_N3, G/LMO_Ag3_P3, G/LMO_Ag3_DMF_N3, and G/LMO_Ag3_DMF_P3. (b) Linear voltammograms recorded in a 0.1 M KOH solution using electrodes G/LMO_Ag3, G/LMO_Ag3_N3, and G/LMO_Ag3_N5.

The voltammograms presented in fig. 4-8 a) show the similarity of the overpotential values of the modified electrodes at which oxygen is released. The electrocatalytic activity of the G/LMO_Ag3_DMF_P5 electrode is the lowest, while the best performing electrodes are G/LMO_Ag3, G/LMO_Ag3_N3, and G/LMO_Ag3_N5. Because the polarization curves for the three high-performance electrodes overlap to a large extent (fig. 5 b)), identifying the best electrode to catalyze the oxygen release reaction is difficult. Up to current densities of ~20 mA/cm^2, the O_2 release at the G/LMO_Ag3 electrode occurs at higher overpotentials in comparison to the other two electrodes, which means that they have superior electrocatalytic properties. A relevant difference regarding the electrocatalytic properties of the electrodes G/LMO_Ag3_N3 and G/LMO_Ag3_N5 is the fact that, at current densities of > 22 mA/cm^2 in the case of the electrode G/LMO_Ag3_N3, oxygen is released more strongly. In order for water electrolysis to proceed efficiently, it is important to diminish the impact of the different types of voltage drops that can affect this process. One type of voltage drop is due to oxygen bubbles that form during its release and remain attached to the electrode's surface, blocking the active areas of the catalysts and preventing ion transport [78]. The recorded voltammogram shape for the G/LMO_Ag3_N3 electrode indicates the intense release of oxygen bubbles and, implicitly, the pronounced effect of the voltage drop due to these bubbles, clogging the surface of the electrode. The effect is less pronounced in the case of G/LMO_Ag3_N5, given the obtained polarization curve using this electrode. Based on this consideration, the electrode G/LMO_Ag3_N5 was chosen as the electrode with the best electrocatalytic properties for the oxygen release reaction.

The linear voltammograms shown in fig. 6 were plotted for the electrodes obtained by nickel substrate modification containing $LaMnO_3$ perovskite doped with 0.15 % Ag (applied as such or in compositions with Nafion and PEDOT) from suspensions in which the dispersion medium was either ethanol, or DMF.

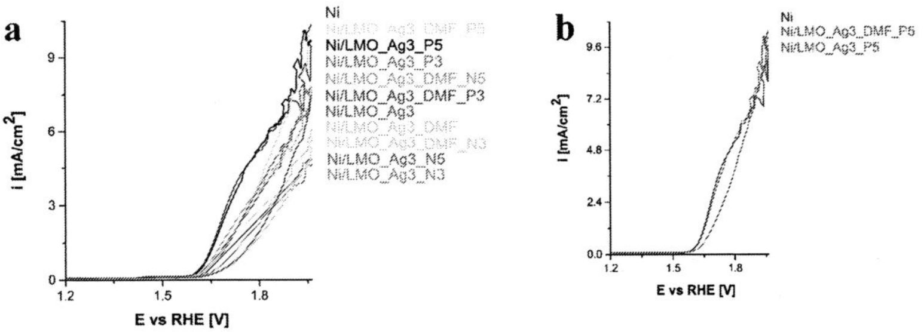

Fig. 4-9. (a) Linear voltammograms recorded in a KOH solution with a concentration of 0.1 M at v = 1 mV/s, using the unmodified Ni electrode (Ni) and the following modified Ni electrodes: Ni/LMO_Ag3, Ni/LMO_Ag3_DMF, Ni/LMO_Ag3_N5, Ni/LMO_Ag3_P5, Ni/LMO_Ag3_DMF_N5, Ni/LMO_Ag3_DMF_P5, Ni/LMO_Ag3_N3, Ni/LMO_Ag3_P3, Ni/LMO_Ag3_DMF_N3, and Ni/LMO_Ag3_DMF_P3. (b) Linear voltammograms recorded in a 0.1 M KOH solution using electrodes Ni, Ni/LMO_Ag3_DMF_P5, and Ni/LMO_Ag3_P5.

Due to the strong release of oxygen when using the Ni substrate, the potential range in which the voltammograms shown in fig. 6 were recorded is narrower than the curves presented in fig. 5. Another difference has to do with the overpotential at which oxygen is released on the modified electrodes, which decreases significantly when the conductive graphite substrate is replaced with the Ni substrate. This effect is also observed in the case of curves plotted for the unmodified Ni electrode (fig. 6 a)). In consequence, this means that the electrocatalytic activity of the electrode is almost as good as that of the most efficient modified electrode, i.e. the Ni/LMO_Ag3_P5 electrode (fig. 6 b)). The electrocatalytic activity of the Ni/LMO_Ag3_P5 electrode is not significantly better than that of the unmodified Ni electrode, but in the case of the G/LMO_Ag3_N5 electrode, a clear difference was observed in its electrocatalytic properties compared to those of the unmodified graphite electrode. For this reason, between the two modified electrodes, the G/LMO_Ag3_N5 electrode was chosen for further investigation of electrocatalytic performance in relation to the oxygen release reaction.

This electrode was further studied in a KOH solution at concentrations of 0.5 and 1 M. This was done with the knowledge that with an increasing concentration of this electrolyte solution, a decrease in the value of the oxygen overload potential takes place, along with an increase in the maximum value of the current density recorded in the potential range [79]. The linear voltammograms for the electrode G/LMO_ Ag3_N5 in the KOH solution with different concentrations are shown in fig. 7 a), while fig. 7 b) shows the chronoamperometric response of the same electrode at the overpotential value of 0.893 V (corresponding to i = 20 mA/cm^2 on the recorded curve in 1 M KOH), together with the voltammograms obtained on the electrode before and after the chronoamperometric test.

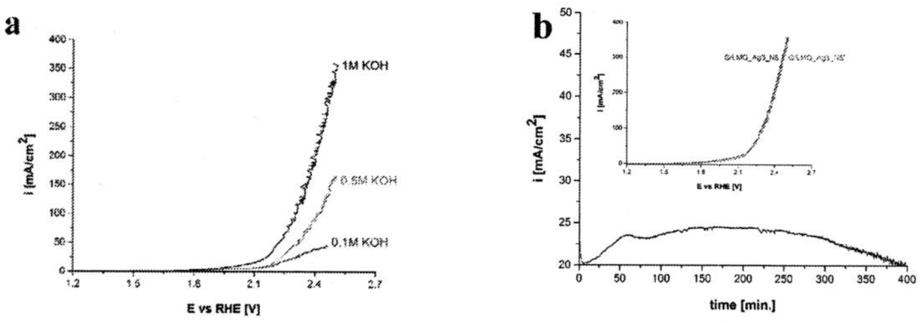

Fig. 4-10. (a) Linear voltammograms recorded on the electrode G/LMO_ Ag3_N5 in a KOH solution with concentrations of 0.1 M, 0.5 M, and 1 M at v = 1 mV/s. (b) The chronoamperometric response of the electrode G/LMO_ Ag3_N5 to the superpower value corresponding to the current density i = 20 mA/cm^2 on the curve recorded for 1 M KOH. Respectively, G/LMO_ Ag3_N5, and G/LMO_ Ag3_N5 ' in a KOH solution with a concentration of 1 M at v = 1 mV/s.

As can be seen in fig. 4-10 a), for the G/LMO_ Ag3_N5 electrode, with an increase in the KCl concentration, we see an increase in the maximum value of the current density and a decrease in the value of the oxygen release overpotential. Because electrocatalytic stability is a parameter for evaluating electrocatalytic activity [80], the G/LMO_Ag3_N5 electrode was subjected to a 400 minute chronoamperometric test to determine its stability. Based on the obtained curve (fig. 4-10 b)), certain statements can be made. Thus, the current density increases from 20 mA/cm^2 to the maximum value of 24.5 mA/cm^2, reached after 172.5 minutes (an increase of 22.5 %). Then, the current density gradually decreases until the end of the test, when it reaches the value of 20 mA/cm^2. The decrease can be attributed to the formation of O_2 bubbles remaining on the surface of the electrode, which makes it difficult to carry out the oxygen release reaction [81]. Fig. 7 b) presents the voltammograms of the electrode before and after the stability test. We can observe that the two curves almost overlap, which means that the modified electrode has retained its original properties.

Another important parameter that indicates the electrocatalytic performance of catalysts is the Tafel slope. Its value, which is intended to be as small as possible [82, 83], counts in the study of the kinetics of the oxygen release reaction at the electrode/electrolyte interface. To highlight a kinetic effect and not just the one related to surface, the current density was expressed as a function of the electroactive surface of the G/LMO_ Ag3_N5 electrode. This was estimated using the Randles-Sevcik equation and the data obtained after recording cyclic voltammograms, at different scanning speeds, in the electrolyte solution composed of KNO3 at a 1 M concentration and K3 [Fe (CN) 6] with a 4 mM concentration. The voltammograms shown in fig. 8 a) present the anodic and cathodic peaks corresponding to the ferric/ferrocyanide redox couple increase with increasing scanning speed. The dependence between the peak current densities of the two peaks and the scan rate radical is shown in fig. 8 b). This shows that peak currents are proportional to the scanning speed radical and this the behavior is characteristic of a diffusion-controlled electronic transfer process [84].

The values for the electroactive surface of the electrode G/LMO_Ag3_N5, as well as for the diffusion coefficient of ferricyanide ions, estimated using the Randles-Sevcik equation, are: 0.68 cm^2, respectively 4.35 x 10-5 cm^2/s. A linear voltammogram recorded on the electrode G/LMO_Ag3_N5, in KOH with a concentration of 1 M, for which the current density was expressed as a function of the electroactive surface of the electrode, is shown in fig. 8 c). In addition, this figure presents the linear region of the Tafel curve obtained for the process of oxygen release on the modified electrode.

Using the Tafel equation (equation 4), we obtained a Tafel slope value of 0.33 V/dec:

$$\eta = b\, x \log(i) + a \qquad (4.4)$$

where η is the oxygen overload potential [V]; i is the current density [mA/cm^2]; and b is the Tafel slope.

In conclusion, the study of electrocatalytic activity for the oxygen release reaction of electrodes modified with undoped/doped perovskite material with metal ions (Ag and Pd) in different concentrations, shows that the graphite electrode modified by an ethanol suspension containing $LaMnO_3$ doped with 0.15 % Ag and Nafion has the best overall electrocatalytic properties. In-depth investigation of these properties has shown that, for the respective electrode the oxygen overload potential at igeom = 20 mA/cm^2 is 0.893 V, keeping its electrocatalytic properties after the stability test and corresponding to a Tafel slope value of 0.33 V/dec.

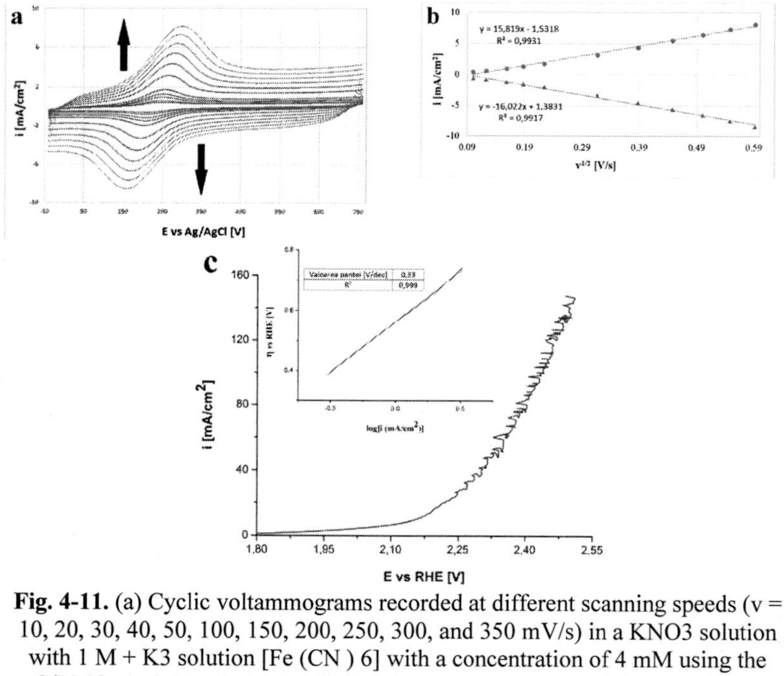

Fig. 4-11. (a) Cyclic voltammograms recorded at different scanning speeds (v = 10, 20, 30, 40, 50, 100, 150, 200, 250, 300, and 350 mV/s) in a KNO3 solution with 1 M + K3 solution [Fe (CN) 6] with a concentration of 4 mM using the G/LMO_Ag3_N5 electrode. (b) Graphical representation of the dependence between the peak current densities of the anodic and cathodic peaks corresponding to the ferrox/ferrocyanide redox torque and the scanning speed radical. (c) Linear voltamogram drawn for the electrode G/LMO_Ag3_N5 in a KOH solution with a concentration of 1 M at v = 1 mV/s for which the current density was expressed as a function of the electroactive surface of the electrode. The linear region of the Tafel curve obtained in the case of the oxygen release reaction on the modified electrode is also inserted in the figure.

4.3.2.2. Investigation of the electrocatalytic activity of electrodes modified with perovskites for the oxygen reduction reaction

Graphite electrodes modified with perovskites, as described in procedures 1-4 and listed in table 4-1 (presented in the experimental part), were the subject of a preliminary study regarding their electrocatalytic activity for the oxygen reduction reaction. These electrochemical experiments produced cyclic voltammogram recordings in an alkaline medium through which O_2 was bubbled. The appearance of the voltammograms for the modified electrodes with undoped and Ag-doped $LaMnO_3$ was compared to the curves obtained by Xue et al. [52] under similar conditions. The peak potential values corresponding to the oxygen reduction peak for the modified electrodes are shown in table 4-6, while the cyclic voltammograms of the electrodes modified following the aforementioned four procedures are shown in fig. 9 (for procedures 1 and 2) and in fig.10 (for procedures 3 and 4).

Table 4-6. Peak potential values corresponding to the oxygen reduction peak for electrodes modified by procedures 1-4.

Procedure 1		***Procedure 2***		***Procedure 3***		***Procedure 4***	
Electrode name	*E vs RHE [V]*	*Electrode name*	*E vs RHE [V]*	*Electrode name*	*E vs RHE [V]*	*Electrode name*	*E vs RHE [V]*
G/LMO	0,38	G/LMO_N	-*	G/LMO_V	0,519	G/LMO_V_N	0,3365
G/LMO_Ag1	0,36	G/LMO_Ag1_N	0,3326	G/LMO_Ag1_V	0,47	G/LMO_Ag1_V_N	0,352
G/LMO_Ag2	0,363	G/LMO_Ag2_N	0,432	G/LMO_Ag2_V	0,533	G/LMO_Ag2_V_N	0,3911
G/LMO_Ag3	0,336	G/LMO_Ag3_N	0,4652	G/LMO_Ag3_V	0,37	G/LMO_Ag3_V_N	0,4178
G/LMO_Pd1	0,4515	G/LMO_Pd1_N	0,4044	G/LMO_Pd1_V	0,31	G/LMO_Pd1_V_N	0,3336
G/LMO_Pd2	0,393	G/LMO_Pd2_N	0,457	G/LMO_Pd2_V	0,51	G/LMO_Pd2_V_N	0,4868
G/LMO_Pd3	0,43	G/LMO_Pd3_N	0,454	G/LMO_Pd3_V	0,493	G/LMO_Pd3_V_N	0,494

Better electrocatalytic activity for the oxygen reduction reaction is obtained with a more positive value for the peak potential corresponding to the oxygen reduction process [52]. The results in table 5 indicate that of the perovskite-modified electrodes using procedure 1, the G/LMO_Pd1 electrode has the highest electrocatalytic activity for the oxygen reduction reaction. The same is valid for the electrodes G/LMO_Ag3_N, G/LMO_Ag2_V, and G/LMO_Pd3_V_N, obtained by procedures 2, 3, and 4, respectively. Among all the investigated perovskite-modified electrodes, the G/LMO_Ag2_V electrode was found to be the most promising in terms of electrocatalytic properties for the oxygen reduction reaction. This observation was also the conclusion of the preliminary study on the electrocatalytic activity of the modified electrodes in the case of the oxygen reduction reaction.

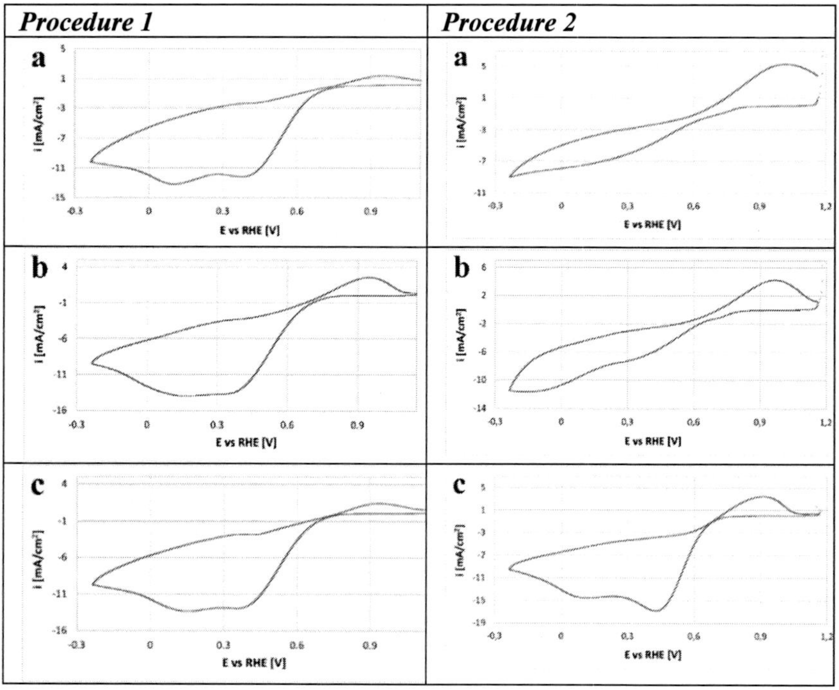

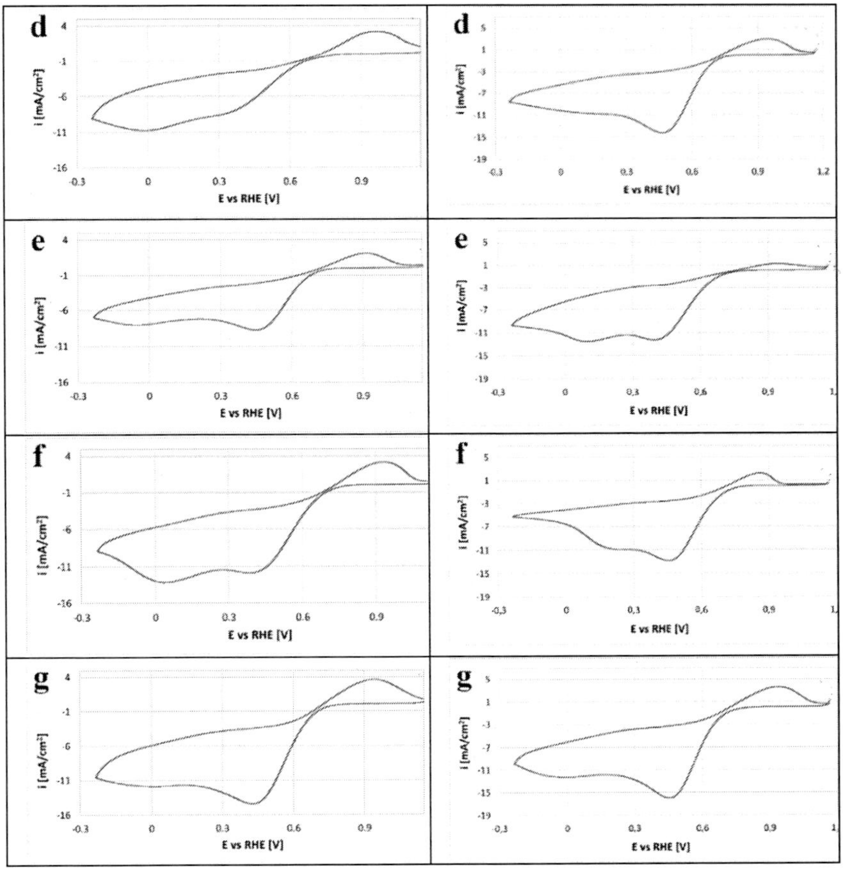

Fig. 4-12. Cyclic voltammograms recorded in a KOH solution with a concentration of 0.1 M at v = 5 mV/s using graphite electrodes modified by procedures 1 and 2. Electrodes modified by procedure 1: G/LMO (a), G/LMO_Ag1 (b), G/LMO_Ag2 (c), G/LMO_Ag3 (d), G/LMO_Pd1 (e), G/LMO_Pd2 (f), and G/LMO_Pd3 (g). Electrodes modified by procedure 2: G/LMO_N (a), G/LMO_Ag1_N (b), G/LMO_Ag2_N (c), G/LMO_Ag3_N (d), G/LMO_Pd1_N (e), G/LMO_Pd2_N (f), and G/LMO_Pd3_N (g).

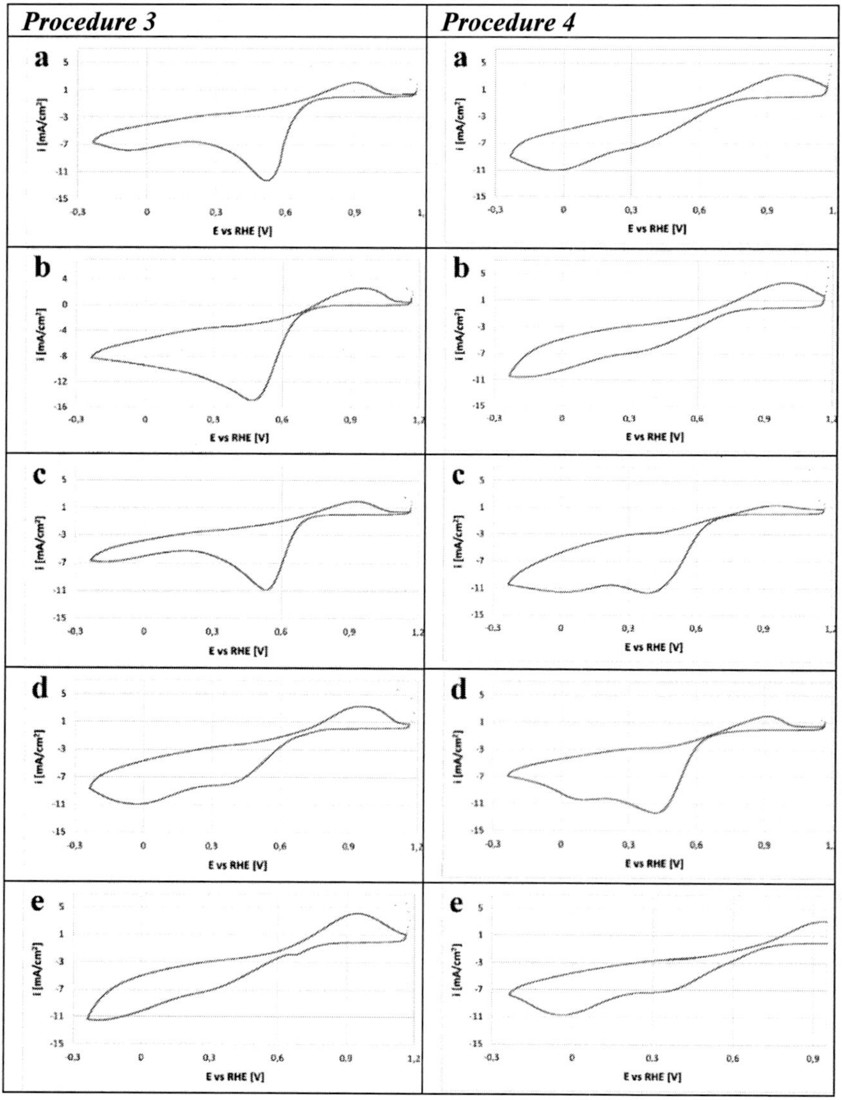
Procedure 3
Procedure 4
a
b
c
d
e
i [mA/cm²]
E vs RHE [V]
5
1
-3
-7
-11
-15
4
0
-4
-8
-12
-16
-0,3
0
0,3
0,6
0,9
1,2

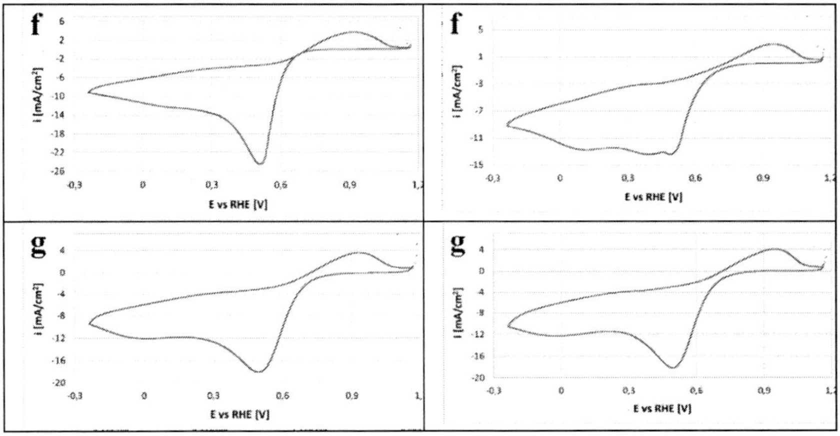

Fig. 4-13. Cyclic voltammograms recorded in a KOH solution with a concentration of 0.1 M at v = 5 mV/s using graphite electrodes modified by procedures 3 and 4. Electrodes modified by procedure 3: G/LMO_V (a), G/LMO_Ag1_V (b), G/LMO_Ag2_V (c), G/LMO_Ag3_V (d), G/LMO_Pd1_V (e), G/LMO_Pd2_V (f), and G/LMO_Pd3_V (g). Electrodes modified by procedure 4: G/LMO_V_N (a), G/LMO_Ag1_V_N (b), G/LMO_Ag2_V_N (c), G/LMO_Ag3_V_N (d), G/LMO_Pd1_V_N (e), G/LMO_Pd2_V_N (f), and G/LMO_Pd3_V_N (g).

4.3.2.3. Investigation of the electrocatalytic activity of electrodes modified with perovskites for the oxidation reaction of nitrite

The study of the electrocatalytic activity of modified electrodes with perovskites for the oxidation reaction of nitrite was performed with procedures 5-7 for the electrodes named according to table 3 (the table and the procedures are in the experimental part) and tracing cyclic voltammograms in the absence and presence of different concentrations of $NaNO_2$, in the range 1.457 x 10^{-4} to 45.19 x 10^{-4} M.

Figure 4-14 a) shows the curves recorded using the unmodified graphite electrode in the presence of $NaNO_2$ concentrations in the range 4.353 x 10^{-4} to 45.19 x 10^{-4} M. Figure 11 graphically represents the dependence between the peak current density of the anodic signal attributed to nitrous oxide oxidation and the concentration of $NaNO_2$ in the electrolyte solution for the concentration range 7.227 x 10^{-4} to 45.19 x 10^{-4} M.

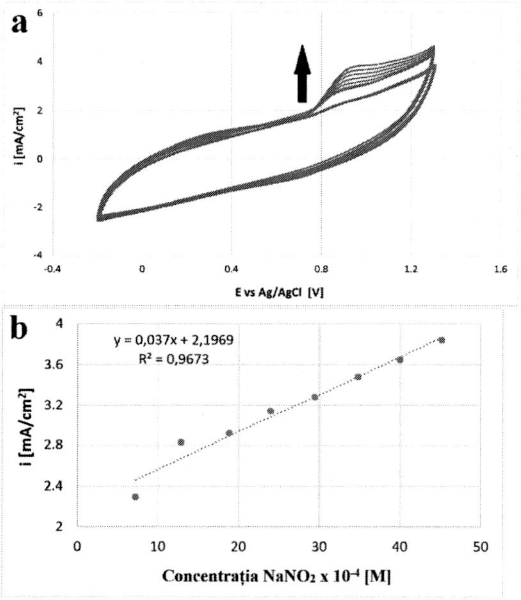

Fig. 4-14. (a) Cyclic voltammograms recorded using the unmodified graphite electrode in a KCl solution with a concentration of 0.1 M in the presence of $NaNO_2$ concentrations in the range 4.353 x 10^{-4} to 45.19 x 10^{-4} M, la v = 100 mV/s. (b) Graphical representation of the dependence between the peak current density of the anodic signal corresponding to the oxidation of nitrite and the concentration of $NaNO_2$ in the electrolyte solution for the concentration range 7.227 x 10^{-4} to 45.19 x 10^{-4} M.

For the voltammograms of the unmodified electrode in the presence of 10 and 20 μL $NaNO_2$ added to the KCl solution, no signal was observed that could be attributed to the nitrite oxidation process. When adding volumes of 30, 40, and 50 μL $NaNO_2$ to the KCl solution, the recorded voltammograms are seen to overlap at the potential of 0.88 V, which represents the peak potential for a weak signal due to nitrous oxide. By increasing the nitrite concentration, the peak potential of the signal shifts to more positive values (from 0.88 to 0.97 V), which is a behavior commonly observed in the amperometric study of electrodes with sensory properties [85, 86].

In addition to the displacement of the potential, the increase in the nitrite concentration also has the effect of intensifying the anodic signal (fig. 4-14 b)), justifying its attribution to the nitrite oxidation reaction. In addition, in the nitrite oxidation experiments conducted by Kozub et al.

[37] on an unchanged vitreous carbon electrode, the signal attributed to the nitrite oxidation reaction at 2 mMa was observed around a potential of 0.8 V vs. SCE, or 0.844 V vs. AgAgCl (sat. KCl). This value is close to that at which the anodic signal for the unmodified graphite electrode appears.

In the case of the dependence between the peak current density of the anodic signal attributed to nitrous oxide oxidation and the $NaNO_2$ concentration in the electrolyte solution (fig. 4-14 b)) the following linear regression equation was obtained for the concentration range 7.227 x 10^{-4} to 45.19 x 10^{-4} M: i [mA/cm^2] = 0.037 x concentration $NaNO_2$ x 10^{-4} [M] + 2.1969 and R^2 = 0.9673. The cyclic voltammograms for the G/LMO_DMF electrode (modified by procedure 5) in the presence of $NaNO_2$ concentrations in the range 4.353 x 10^{-4} to 45.19 x 10^{-4} M are presented in fig. 4-15 a)). Figure 4-15 b) shows the dependence between the peak current density of the anodic signal attributed to nitrite oxidation and the concentration of $NaNO_2$ in the electrolyte solution.

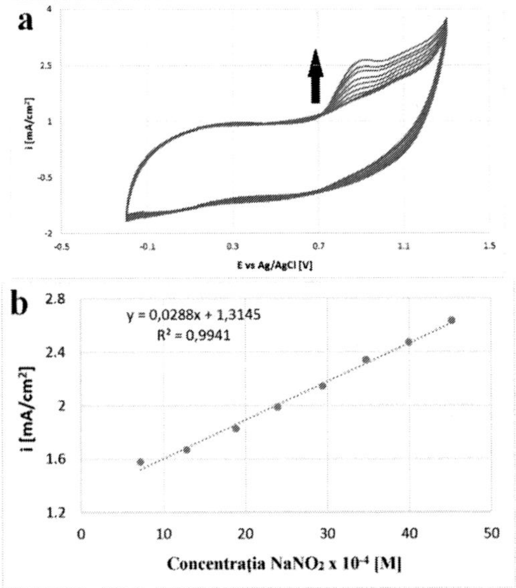

Fig. 4-15. (a) Cyclic voltammograms recorded using the G/LMO_DMF electrode in a KCl solution with a concentration of 0.1 M in the presence of $NaNO_2$ concentrations in the range 4.353 x 10^{-4} to 45.19 x 10^{-4} M at v = 100 mV/s. (b) Graphical representation of the dependence between the peak current density of the anodic signal corresponding to nitrous oxide oxidation and the concentration of $NaNO_2$ in the electrolyte solution for the concentration range 7.227 x 10^{-4} to 45.19 x 10^{-4} M.

The electrochemical behavior of the G/LMO_DMF electrode in the presence of nitrite is similar to that observed for the unmodified electrode. For volumes of 10 and 20 μL $NaNO_2$ added to the KCl solution, no signal was observed that could be attributed to the nitrite oxidation process. For volumes of 30, 40, and 50 μL $NaNO_2$ the voltammograms overlap at a potential of 0.86 V, which represents the peak potential for a weak signal highlighted by nitrite oxidation. With an increasing concentration of nitrite in the electrolyte solution, the signal increases in intensity and moves to more positive values (from 0.86 to 0.91 V), as shown in fig. 4-15 a). We should also mention that in nitrite ion detection studies using electrodes modified with perovskitic materials reported by Yatsunami et al. [4], nitrite oxidation took place around a potential of 0.85 V vs. Ag/AgCl (sat. KCl). For the dependence between the peak current density of the anodic signal attributed to nitrous oxide oxidation and the $NaNO_2$ concentration in the electrolyte solution (fig. 4-15 b)), the following linear regression equation was obtained: i [mA/cm^2] = 0.0288 x $NaNO_2$ concentration x 10-4 [M] + 1.3145 and R^2 = 0.9941.

The voltammograms shown in fig. 13 a) present the recorded electrochemical responses of the G/LMO_DMF_V electrode (modified by procedure 6) to different nitrite concentrations. For this electrode, as for those previously presented, the dependence between the peak current density of the anodic signal corresponding to the nitrous oxide oxidation and concentration of $NaNO_2$ in the electrolyte solution was highlighted (fig. 13 b)). Unlike the unmodified electrode and G/LMO_DMF, the G/LMO_DMF_V electrode proved to be more sensitive to nitrite. Thus, the voltammograms recorded for volumes of 30, 40, and 50 μL $NaNO_2$ in the KCl solution did not overlap with the peak potential of the signal attributed to the nitrite oxidation process. For this electrode, it was present at 0.92 V (for a concentration of 4.353 x 10-4 M $NaNO_2$). Figure 4-16 a) shows that, even in the case of this electrode, there is a shift towards more positive values (from 0.92 to 0.98 V) and an increase in signal intensity corresponding to the oxidation of nitrite as its concentration increases in the electrolyte solution.

The linear regression equation obtained in the case of the dependence between the peak current density of the anodic signal attributed to nitrous oxide and the concentration of $NaNO_2$ in the electrolyte solution (fig. 13 b)) is: i [mA/cm^2] = 0.0411 x $NaNO_2$ concentration x 10-4 [M] + 5.9415 and R^2 = 0.9992.

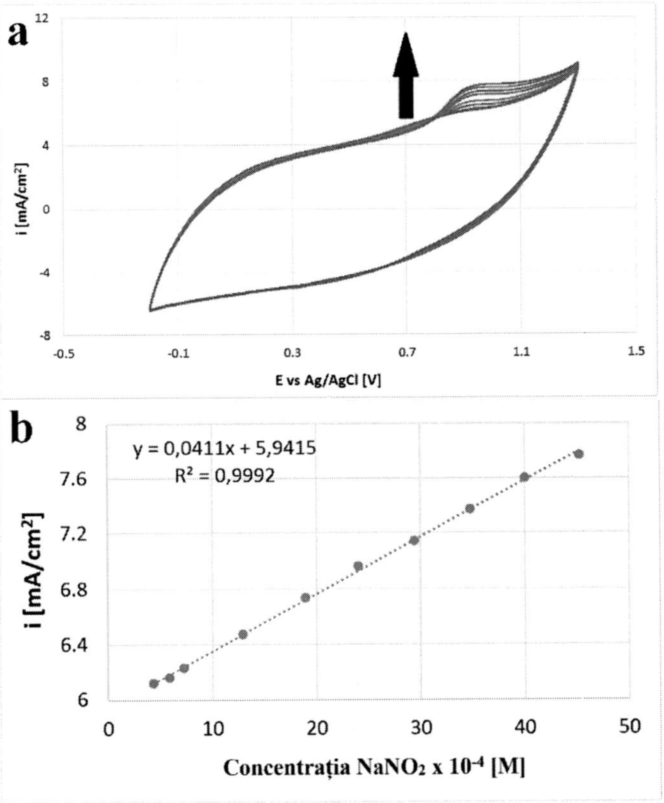

Fig. 4-16. (a) Cyclic voltammograms recorded using the G/LMO_DMF_V electrode in a KCl solution with a concentration of 0.1 M in the presence of $NaNO_2$ concentrations in the range 4.353 x 10^{-4} to 45.19 x 10^{-4} M at v = 100 mV/s. (b) Graphical representation of the dependence between the peak current density of the anodic signal attributed to nitrous oxide and the concentration of $NaNO_2$ in the electrolyte solution for the concentration range 4.353 x 10^{-4} to 45.19 x 10^{-4} M.

Figure 4-16 a) shows the cyclic voltammograms obtained for the G/LMO_DMF_V_N electrode (modified by procedure 7) in the presence of $NaNO_2$ concentrations in the range 4.353 x 10-4 to 45.19 x 10-4 M. Fig. 14 b) shows the dependence between the peak current density of the anodic signal corresponding to the oxidation of nitrite and the concentration of $NaNO_2$ in the electrolyte solution is observed.

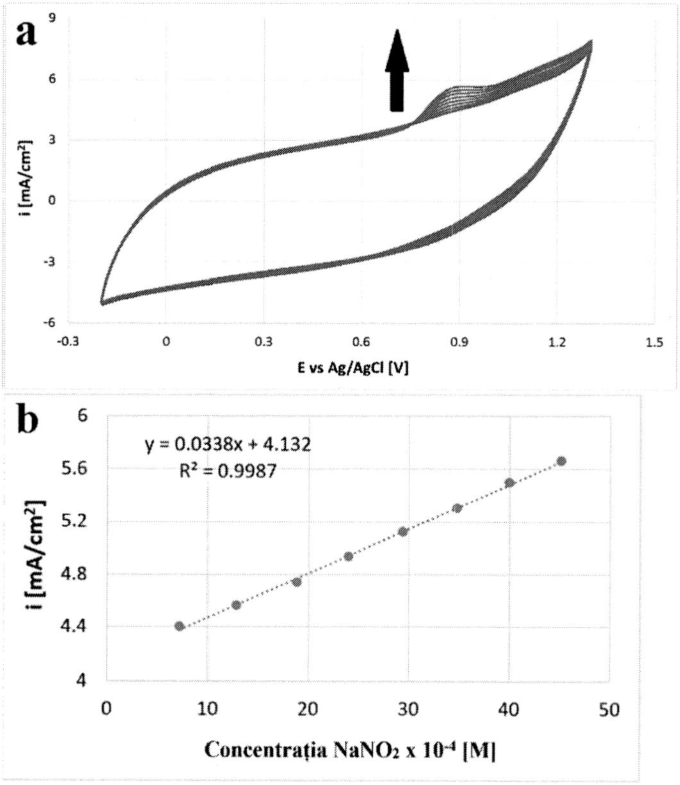

Fig. 4-17. (a) Cyclic voltammograms recorded using the G/LMO_DMF_V_N electrpode in a KCl solution with a concentration of 0.1 M,in the presence of $NaNO_2$ whit concentrations in the range $4.353 \cdot 10^{-4}$ to $45.19 \cdot 10^{-4}$ M at v = 100 mV/s. (b) Graphical representation of the dependence between the peak current density of the anodic signal attributed to nitrous oxide and the concentration of $NaNO_2$ in the electrolyte solution for the concentration range $7.227 \cdot 10^{-4}$ to $45.19 \cdot 10^{-4}$ M.

The behavior of the electrode G/LMO_DMF_V_N in the presence of nitrite is similar to that of the unmodified electrode and that of G/LMO_DMF. A signal attributable to nitrite oxidation was observed when a volume of 30 μL $NaNO_2$ was added to the KCl solution and overlapped the signals identified on the curves drawn after the addition of volumes of 40 and 50 μL $NaNO_2$. The increase in the nitrite concentration in the electrolyte solution leads to the intensification of the signal that can be attributed to nitrite oxidation and its displacement towards more positive potentials (fig. 4-17 a) from 0.87 to 0.91 V.

As can be seen in fig. 14 b), the linear regression equation obtained for the dependence between the peak current density of the anodic signal attributed to nitrous oxide and the $NaNO_2$ concentration in the electrolyte solution is: i [mA/cm^2] = 0.0338 x $NaNO_2$ concentration x 10-4 [M] + 4.132 and $R^2 = 0.9987$.

From the investigated electrodes as regards their electrocatalytic activity for the oxidation reaction of nitrite (G, G/LMO_DMF, G/LMO_DMF_V, and G/LMO_DMF_V_N), it was observed that the electrode G/LMO_DMF_V can catalyze this reaction if the condition of nitrite from the electrolyte solution is 4.353 x 10-4 M showing a higher sensitivity than the other electrodes studied.

In order to obtain an electrode with even higher sensitivity to nitrite detection, modified electrodes were made with $LaMnO_3$ doped with 0.15 % Pd and 0.15 % Ag (G/LMO_Pd3_DMF_V and G/LMO_Ag3_DMF_V, respectively). The change was made by procedure 6, given that the electrode G/LMO_DMF_V was obtained by the same procedure. Figures 15 and 16 show the experimental results obtained using the two modified electrodes with doped perovskite materials.

The nitrite sensitivity of the G/LMO_Pd3_DMF_V electrode was found to be higher than that of the G/LMO_DMF_V electrode. Thus, on the curve drawn in the KCl solution in which 20 μL $NaNO_2$ was added, a weak signal was observed, the intensity of which increased with increasing nitrite concentration (fig. 15 a)). Its peak potential moved in the concentration range 2.908 x 10-4 to 45.19 x 10-4 M from 0.855 to 0.915 V. In the case of the dependence between the peak current density of the anodic signal attributed to nitrite oxidation and the concentration of $NaNO_2$ from the electrolyte solution (fig. 15 b)) the following linear regression equation was obtained: i [mA/cm^2] = 0.0531 x $NaNO_2$ concentration x 10-4 [M] + 3.7381, and $R^2 = 0.999$.

The best sensitivity was observed in the case of the G/LMO_Ag3_DMF_V electrode, which responded to the analyte with a volume of 10 μL $NaNO_2$ added to the KCl solution. As the concentration of nitrite in the electrolyte solution increased, the signal intensity increased (fig. 16 a)) and its peak potential moved in the concentration range 1.457 x 10-4 to 45.19 x 10-4 M from 0.84 to 0.88 V. The potential values indicate that nitrite oxidation at this electrode occurs at lower potentials compared to G/LMO_DMF_V and G/LMO_Pd3_DMF_V electrodes. For example, at a concentration of 4.353 x 10-4 M $NaNO_2$ the peak potential of the signal attributed to nitrous oxide is 0.92 V for

G/LMO_DMF_V; 0.866 V for G/LMO_Pd3_DMF_V; and 0.85 V for G/LMO_Ag3_DMF_V. Moreover, because the oxidation potential is shifted to lower values, the electrocatalytic activity of G/LMO_Ag3_DMF_V is higher than that of the other electrodes obtained by procedure 6 [87].

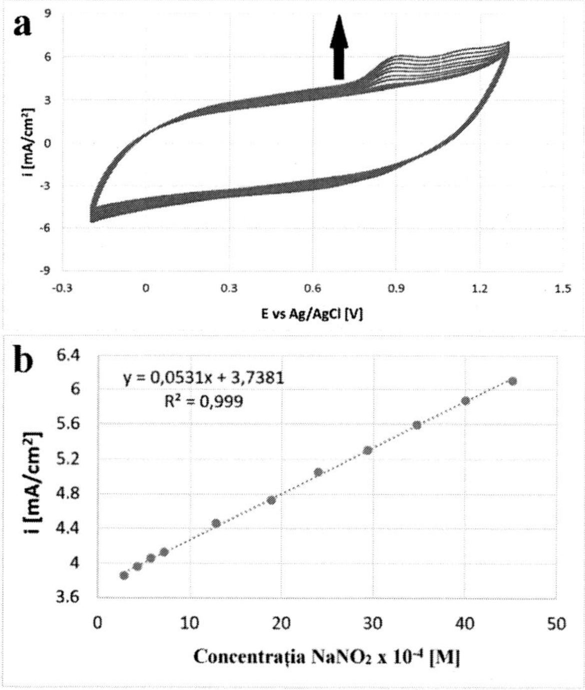

Fig. 4-18. (a) Cyclic voltammograms recorded using the electrode G/LMO_Pd3_DMF_V in KCl solution with a concentration of 0.1 M, in the absence and presence of $NaNO_2$ concentrations in the range 1.457×10^{-4} to 45.19×10^{-4} M at v = 100 mV/s. (b) Graphical representation of the dependence between the peak current density of the anodic signal corresponding to the oxidation of nitrite and the concentration of $NaNO_2$ in the electrolyte solution, for the concentration range 2.908×10^{-4} to 45.19×10^{-4} M.

The superior electrocatalytic properties of the modified Ag-doped $LaMnO_3$ electrode compared to those of the undoped $LaMnO_3$ and Pd-doped $LaMnO_3$ electrodes can be attributed to the existence of a synergistic effect between the materials with which the electrode has been modified [88]. In the present case, this effect is likely to occur through the interaction of Ag-doped perovskite and Vulcan conductive carbon.

Figure 16 b) shows the dependence between the peak current density of the anodic signal corresponding to the oxidation of nitrite observed using the G/LMO_Ag3_DMF_V electrode and the concentration of $NaNO_2$ in the electrolyte solution for the concentration range 1.457 x 10-4 to 45.19 x 10-4 M. The linear regression equation obtained is: i [mA/cm^2] = 0.0514 x $NaNO_2$ concentration x 10-4 [M] + 2.5339 and R^2 = 0.9989.

Looking at the overlap of the voltammograms for the unmodified graphite electrode and G/LMO_Ag3_DMF_V in an electrolyte solution containing 45.19 x 10-4 M $NaNO_2$ (fig. 4-19 c)), we can observe clear differences between the anodic peak attributed to nitrite oxidation on the two polarization curves. The potential at which the anodic peak appears in the case of the modified electrode is lower than that at which the same peak appears for the unmodified electrode. In addition, the intensity of the anodic peak in the case of the unmodified electrode is lower than in the case of the modified one. Both observations indicate the superior electrocatalytic properties of G/LMO_Ag3_DMF_V for the nitrite oxidation reaction.

The G/LMO_Ag3_DMF_V electrode was also electrochemically characterized by recording cyclic voltammograms at different scanning speeds in the presence of a concentration of 45.19 x 10-4 M $NaNO_2$ (fig. 4-19 d)). Based on the results obtained, the dependence between the peak current density of the anodic signal attributed to nitrous oxide oxidation and the square root of the scanning speed has been graphically represented (fig. 16 e)). The linear dependence between the anodic peak current and v1/2, with R^2 = 0.9992, indicates that nitrite oxidation is a diffusion-controlled electron transfer process [89].

The main conclusion from our study of electrocatalytic activity for the oxidation reaction of nitrite, carried out on modified electrodes with undoped and ion metal-doped perovskitic material (Pd and Ag), is that an electrode modified with $LaMnO_3$ doped with 0.15 % Ag displays the best electrocatalytic properties.

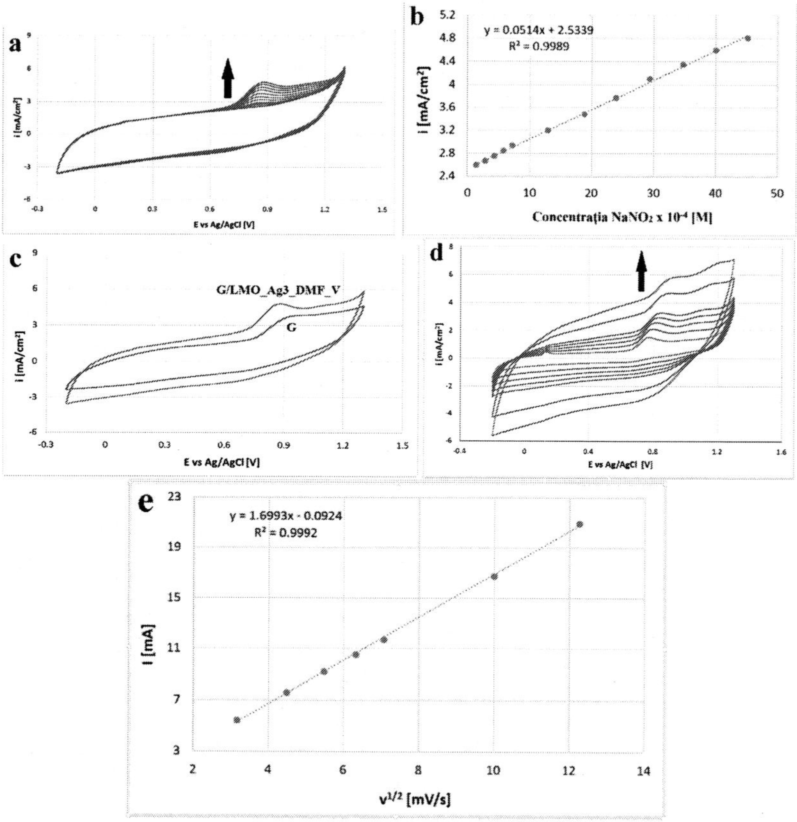

Fig. 4-19. (a) Cyclic voltammograms recorded using the G/LMO_Ag3_DMF_V electrode in a KCl solution with a concentration of 0.1 M in the absence and presence of $NaNO_2$ concentrations in the range 1.457 x 10^{-4} to 45.19 x 10^{-4} M at v = 100 mV/s. (b) Graphical representation of the dependence between the peak current density of the anodic signal corresponding to nitrous oxide oxidation and the concentration of $NaNO_2$ in the electrolyte solution for the concentration range 1.457 x 10^{-4} to 45.19 x 10^{-4} M. (c) Cyclic voltammograms recorded on the unmodified graphite electrode (G) and on G/LMO_Ag3_DMF_V in KCl solutions with a concentration of 0.1 M in the presence of $NaNO_2$ (45.19 x 10^{-4} M). (d) Cyclic voltammograms recorded at different scanning speeds (v = 10, 20, 30, 40, 50, 100, and 150 mV/s) in a 0.1 M KOH solution and in the presence of $NaNO_2$ (45.19 x 10^{-4} M) using the G/LMO_Ag3_DMF_V electrode. (e) Graphical representation of the dependence between the peak current density of the anodic signal attributed to nitrous oxide oxidation and the square root of the scanning speed.

4.4. Conclusion

In this study, undoped and doped (Ag, Pd, and N) $LaMnO_3$ materials synthesized by the sol-gel method were annealed at 600 °C for 6 h. X-ray diffraction analysis showed that all the materials obtained were well crystallized and corresponded to the perovskite phase.

The three electrochemical studies described in this study have led to the identification of perovskite-modified electrodes with the best electrocatalytic properties for the oxygen release reaction, hydrogen reduction reaction, and nitrite oxidation reaction.

In the case of the oxygen release reaction, the preliminary results obtained on the electrodes modified by four different procedures indicate that the graphite electrode modified with $LaMnO_3$ and doped with 0.15 % Ag has the highest electrocatalytic activity. The optimization study based on these preliminary results used different conductive substrates (graphite and nickel), dispersing agents (ethanol and dimethylformamide), and binders (Nafion and PEDOT). The influence of the substrate, type of dispersing agent, as well as the type and amount of binder on the electrochemical characteristics of the modified electrodes was taken into account. The experimental results highlighted the electrocatalytic properties of perovskitic and hybrid materials (compositions containing materials other than perovskites) used in the manufacture of electrodes. The most effective proved to be the graphite electrode modified by deposition in suspension in ethanol of a composition of $LaMnO_3$ doped with 0.15 % Ag and Nafion. Additional electrochemical investigations of this electrode showed that the oxygen overload potential at the igeom = 20 mA/cm^2 is 0.893 V, the electrode retains its electrocatalytic properties after the stability test and corresponds to a Tafel slope value of 0.33 V/dec.

Regarding the study of the electrocatalytic properties for the oxygen reduction reaction of electrodes modified with perovskitic materials, the preliminary results indicate that the graphite electrode modified by ethanol suspension of a composition of $LaMnO_3$ doped with 0.1 % Ag and conductive carbon is the most efficient.

The investigation of the electrocatalytic activity for the nitrous oxide reaction took place in two stages. In the first stage, the unmodified and modified electrodes with undoped $LaMnO_3$ were studied and the experimental results showed that the graphite electrode modified by the deposition in suspension in DMF of a composition of undoped $LaMnO_3$ and conductive carbon has the highest nitrite sensitivity. It could catalyze the oxidation reaction of nitrite at a concentration of 4.353 x 10-4 M $NaNO_2$ present in the electrolyte solution. In the second stage, the properties of this electrode were compared with those of the electrodes modified by the same procedure, but using $LaMnO_3$ doped with 0.15 % Ag and 0.15 % Pd, respectively. According to the results obtained, the best electrocatalytic properties for the oxidation reaction of nitrite are shown by the graphite electrode modified by deposition in suspension in DMF of a composition of $LaMnO_3$ doped with 0.15 % Ag and conductive carbon.

In view of the conclusions reached in these three studies, it can be stated as a general conclusion that electrodes modified with compositions containing Ag ion-doped perovskites display the best electrocatalytic properties for oxygen oxidation reactions, oxygen reduction reactions, and nitrite oxidation reactions.

References

[1] J. X. Flores-Lasluisa, F. Huerta, D. Cazorla-Amorós, E. Morallón. Structural and morphological alterations induced by cobalt substitution in LaMnO3 perovskites, *Journal of Colloid and Interface Science* 556 (2019) 658−666.

[2] Q. Liu, C. Yang, X. Dong, F. Chen. Perovskite Sr2Fe1. 5Mo0. 5O6-δ as electrode materials for symmetrical solid oxide electrolysis cells, *Int. J. Hydrogen Energy* 35 (2010) 10039–10044.

[3] D. B. Meadowcroft. Low-cost oxygen electrode material, *Nature* 226 (1970) 847–848.

[4] W. B. Li, J. X. Wang, H. Gong. Catalytic combustion of VOCs on non-noble metal catalysts, *Catal. Today* 148 (2009) 81–87.

[5] P. Ciambelli, V. Palma, S. F. Tikhov, V. A. Sadykov, L. A. Isupova, L. Lisi. Catalytic activity of powder and monolith perovskites in methane combustion, *Catal. Today* 47 (1999) 199–207.

[6] N. Yamazoe, Y. Teraoka. Oxidation catalysis of perovskites-relationships to bulk structure and composition (valency, defect, etc.), *Catal. Today* 8 (1990) 175–199.

[7] D. K. Hwang, S. Kim, J. H. Lee, I. S. Hwang, I. D. Kim. *J. Mater. Chem.* 21 (2011) 1959–1965.

[8] J. Burschka, N. Pellet, S. J. Moon, R. H. Baker, P. Gao, M. K. Nazeeruddin, M. Grätzel. *Nature* 499 (2013) 316–319.

[9] C. Q. Chen, W. Li, C. Y. Cao, W. G. Song. *J. Mater. Chem.* 20 (2012) 6968–6974.

[10] G. L. Luque, N. F. Ferreyra, A. G. Leyva, G. A. Rivas. *Sens. Actuat. B* 42 (2009) 331–336.

[11] P. Xiaokaiti, T. Yu, A. Yoshida, G. Guan, A. Abudula. Evaluation of cerium doped perovskites (Ce0. 1Sr0. 9) xCo0.3Fe0.7O3-δ as cathode materials for solid oxide fuel cells, *Catalysis Today* 332 (2019) 94–100.

[12] Y. Chen, X. Nie, B. Wang, C. Xia, W. Dong, X. Wang et al. Tuning La0. 6Sr0. 4Co0. 2Fe0. 8O3-δ perovskite cathode as functional electrolytes for advanced lowtemperature SOFCs, *Catalysis Today* (2019).

[13] X. Xu, C. Su, W. Zhou, Y. Zhu, Y. Chen, Z. Shao. Co-doping Strategy for Developing Perovskite Oxides as Highly Efficient Electrocatalysts for Oxygen Evolution Reaction, *Advanced Science* 3 (2016) 1500187.

[14] J. T. Mefford, A. A. Kurilovich, J. Saunders, W. G. Hardin, A. M. Abakumov, R. P. Forslund et al. Decoupling the roles of carbon and metal oxides on the electrocatalytic reduction of oxygen on La 1-x Srx CoO3-δ perovskite composite electrodes, *Physical Chemistry Chemical Physics* 21 (2019) 3327–3338.

[15] L. Zeng, L. Cui, C. Wang, W. Guo, C. Gong. In-situ modified the surface of Ptdoped perovskite catalyst for soot oxidation, *J. Hazard. Mater.* 383 (2020) 121210. https://doi.org/10.1016/j.jhazmat.2019.121210.

[16] M. Li, H. Niu, J. Druce, H. Tellez, T. Ishihara, J. A. Kilner, H. Gasparyan, M. J. Pitcher, W. Xu, J. F. Shin, L. M. Daniels, L. A. H. Jones, V. R. Dhanak, D. Hu, M. Zanella, J. B. Claridge, M. J. Rosseinsky. A CO2-tolerant perovskite

oxide with high oxide ion and electronic conductivity, *Adv. Mater.* 32 (2020) 1905200. https://doi.org/10.1002/adma.201905200.
[17] H. Chang, E. Bjørgum, O. Mihai, J. Yang, H. L. Lein, T. Grande, S. Raaen, Y. A. Zhu, A. Holmen, D. Chen. Effects of oxygen mobility in La-Fe-based perovskites on the catalytic activity and selectivity of methane oxidation, *ACS Catal.* 10 (2020) 3707–3719. https://doi.org/10.1021/acscatal.9b05154.
[18] I. Hussain, S. N. Khan, T. N. Rao, A. Kumar, B. H. Koo. Structural, magnetic and magnetocaloric properties of double perovskites Ba2-xLaxFeMoO6, *Solid State Sci.* 97 (2019) 105991.
https://doi.org/10.1016/j. solidstatesciences.2019.105991.
[19] T. S. Rao, V. R. K. Murthy, B. Viswanathan. Review of Perovskite Ceramics: Microwave Dielectric Resonator Materials 0193 (2011).
https://doi.org/10.108 0/0015019900822147.
[20] Tarancon, A. Strategies for lowering solid oxide fuel cells operating temperature. *Energies* 2 (2009) 1130–50.
https://doi.org/10.3390/en20401130.
[21] P. Holtappels, C. Bagger. *J. Eur. Ceram. Soc.* 22 (2002) 41–48.
[22] E. L. Nagaev, *Phys. Rep.* 346 (2001) 387–531.
[23] A. E. Giannakis, A. K. Ladavos, P. J. Pomonis, *Appl. Catal. B* 49 (2004) 147–158.
[24] Paula Sfirloaga, Iuliana Sebarchievici, Bogdan Taranu, Maria Poienar, Gabriela Vlase, Titus Vlase, Paulina Vlazan. Investigation of physico-chemical features of lanthanum manganitewith nitrogen addition, *Journal of Alloys and Compounds* 843 (2020) 155854.
[25] P. Sfirloaga, C. N. Marin, I. Malaescu, P. Vlazan. The electrical performance of ceramics materials with perovskite structure doped with metallic ions, *Ceramics International* 42 (2016) 18960–18964.
[26] E. Campagnoli, A. Tavaresb, L. Fabbrini, I. Rossetti, Yu. A. Dubitsky, A. Zaopo, L. Forni. *Appl. Catal. B: Environ.* 55 (2005) 133–139.
[27] A. Civera, G. Negro, S. Specchia, G. Saracco, V. Specchia. *Catal. Today* 100 (2005) 275–281.
[28] 34] Z. Zhang, C. Wu, D. Wang, G. Liu, Q. Zhang, W. Luo, X. Qi, X. Guo, Y. Zhang, Y. Lao, B. Qu, L. Xiao, Z. *Chen. Org. Electron.* 74 (2019) 204–210.
[29] [35] H. Machiba, T. Oku, T. Kishimoto, N. Ueoka, A. Suzuki. *Chem. Phys. Lett.* 730 (2019) 117–123.
[30] [36] Y. Fan, H. Qin, W. Ye, M. Liu, F. Huang, D. Zhong. *Thin Solid Films* 667 (2018) 40–47.
[31] D. Chen, C. Chen, Z. M. Baiyee, Z. Shao, F. Ciucci. Non-stoichiometric oxides as low-cost and highly-efficient oxygen reduction/evolution catalysts for lowtemperature electrochemical devices, *Chem. Rev.* 115 (2015) 9869–9921, https://doi.org/10.1021/acs.chemrev.5b00073.
[32] X. Xu, W. Wang, W. Zhou, Z. Shao. Recent advances in novel nanostructuring methods of perovskite electrocatalysts for energy-related applications, *Small Methods* 2 (2018) 1800071
https://doi.org/10.1002/smtd.201800071.

[33] L. Zhu, R. Ran, M. Tadé, W. Wang, Z. Shao. Perovskite materials in energy storage and conversion, *Asia-Pacific J. Chem. Eng.* 11 (2016) 338–369, https:// doi.org/10.1002/apj.2000.

[34] Z. Yang, J. Zhang, M. C. W. Kintner-meyer, X. Lu, D. Choi, J. P. Lemmon, J. Liu. Electrochemical energy storage for green grid, *Chem. Rev.* 111 (2011) 3577–3613. https://doi.org/10.1021/cr100290v.

[35] C. Cui, L. Gan, M. Heggen, S. Rudi, P. Strasser. Compositional segregation in shaped Pt alloy nanoparticles and their structural behaviour during electrocatalysis, *Nat. Mater.* 12 (2013) 765–771. https://doi.org/10.1038/nmat3668.

[36] J. Du, T. Zhang, F. Cheng, W. Chu, Z. Wu, J. Chen. Nonstoichiometric perovskite CaMnO3-d for oxygen electrocatalysis with high activity, *Inorg. Chem.* 53 (2014) 9106–9114. https://doi.org/10.1021/ic501631h.

[37] J. Hu, Q. Liu, Z. Shi, L. Zhang, H. Huang. LaNiO3-nanorod/graphene composite as an efficient bi-functional catalyst for zinc-air batteries, *RSC Adv.* 6 (2016) 86386–86394. https://doi.org/10.1039/C6RA16610E.

[38] Y. Xu, A. Tsou, Y. Fu, J. Wang, J. H. Tian, R. Yang. Carbon-coated perovskite $BaMnO_3$ porous nanorods with enhanced electrocatalytic perporites for oxygen reduction and oxygen evolution, *Electrochim. Acta.* 174 (2015) 551–556. https://doi.org/10.1016/j.electacta.2015.05.184.

[39] H. A. Gasteiger, S. S. Kocha, B. Sompalli, F. T. Wagner. Activity benchmarks and requirements for Pt, Pt-alloy, and non-Pt oxygen reduction catalysts for PEMFCs, *Appl. Catal. B* 56 (2005) 9–35.

[40] J. O'M. Bockris, T. Otagawa. Mechanism of oxygen evolution on perovskites, *J. Phys. Chem.* 87 (1983) 2960–2971.

[41] J. Suntivich, H. A. Gasteiger, N. Yabuuchi, Y. Shao-Horn. Electrocatalytic measurement methodology of oxide catalysts using a thin-film rotating disk electrode, *J. Electrochem. Soc.* 157 (2010) B1263–B1268.

[42] J. B. Goodenough. Electronic and ionic transport properties and other physical aspects of perovskites, *Rep. Prog. Phys.* 67 (2004) 1915–1993.

[43] J. Suntivich, H. A. Gasteiger, N. Yabuuchi, H. Nakanishi, J. B. Goodenough, Y. Shao-Horn. Design principles for oxygen-reduction activity on perovskite oxide catalysts for fuel cells and metal-air batteries, *Nat. Chem.* 3 (2011) 546–550.

[44] F. Gao, R. A. Lewis, X. L. Wang, S. X. Dou. Infrared absorption of lanthanum manganites, *Phys. C. Supercond. Appl.* 341–348 (IV) (2000) 2235–2236. https://doi. org/10.1016/S0921-4534(00)01056-X.

[45] B. M. Nagabhushana, R. P. S. Chakradhar, K. P. Ramesh, C. Shivakumara, G. T. Chandrappa. Combustion synthesis, characterization and metal-insulator transition studies of nanocrystalline La1-xCaxMnO3 ($0.0 \leq x \leq 0.5$), *Mater. Chem. Phys.* 102 (2007) 47–52.
https://doi.org/10.1016/j.matchemphys.2006.11.002.

[46] Mohamed S. Afify, Mohamed M. El Faham, Usama Eldemerdash, Waleed M. A. El Rouby, S. I. El-Dek. Room temperature ferromagnetism in Ag-doped LaMnO3 nanoparticles, *Journal of Alloys and Compounds* 861, 25 April 2021, 158570.

[47] G. Busca, V. Lorenzelli. Infrared spectroscopic identification of species arising from reactive adsorption of carbon oxides on metal oxide surfaces. *J. Mater. Chem.* 7 (1982) 89−126.
[48] V. S. Kolat, H. Gencer, M. Gunes, S. Atalay. Effect of B-doping on the structural, magnetotransport and magnetocaloric properties of La0.67Ca0.33MnO3 compounds, *Mater. Sci. Eng. B Solid-State Mater. Adv. Technol.* 140 (2007) 212–217 https://doi.org/10.1016/j.mseb.2007.05.002.
[49] R. P. S. Chakradhar, B. M. Nagabhushana, G. T. Chandrappa, K. P. Ramesh, J. L. Rao. Solution combustion derived nanocrystalline macroporous wollastonite ceramics, *Mater. Chem. Phys.* 95 (2006) 169–175. https://doi.org/10.1016/j. matchemphys.2005.06.002.
[50] J. Hwang, R. R. Rao, L. Giordano, Y. Katayama, Y. Yu, Y. Shao-Horn. Perovskites in catalysis and electrocatalysis, *Science* 358 (2017) 751–756.
[51] N. Labhasetwar, G. Saravanan, S.K. Megarajan, N. Manwar, R. Khobragade, P. Doggali, F. Grasset. Perovskite-type catalytic materials for environmental applications, Sci. Technol. Adv. Mater. 16 (2015) 036002.
[52] Y. Xue, H. Miao, S. Sun, Q. Wang, S. Li, Z. Liu. $La_{1-x}Ag_xMnO_3$ electrocatalyst with high catalytic activity for oxygen reduction reaction in aluminium air batteries, *RSC Adv.* 7 (2017) 5214-5221.
[53] T. Yatsunami, S. Takase, Y. Shimizu. Amperometric nitrite-ion sensor based on electrodeposited Sm-based perovskite-type oxide thick-film electrode, *Sens. Mater.* 28 (2016) 777–784.
[54] H. Chen, G. Wei, X. Liang, P. Liu, H. He, Y. Xi, J. Zhu. The distinct effects of substitution and deposition of Ag in perovskite $LaCoO_3$ on the thermally catalytic oxidation of toluene, *Appl. Surf. Sci.* 489 (2019) 905–912.
[55] K. Zhou, H. Chen, Q. Tian, Z. Hao, D. Shen, X. Xu. Pd-containing perovskite-type oxides used for three-way catalysts, *J. Mol. Catal. A-Chem.* 189 (2002) 225–232.
[56] B. Kucharczyk, W. Tylus. Effect of Pd or Ag additive on the activity and stability of monolithic $LaCoO_3$ perovskites for catalytic combustion of methane, *Catal. Today* 90 (2004) 121–126.
[57] S. Deng, F. Yang, Q. Zhang, Y. Zhong, Y. Zeng, S. Lin, X. Wang, X. Lu, C. Z. Wang, L. Gu, X. Xia, J. Tu. Phase modulation of (1T-2H)-$MoSe_2$/TiC-C shell/core arrays via nitrogen doping for highly efficient hydrogen evolution reaction, *Adv. Mater.* 30 (2018) 1802223.
[58] X. Miao, L. Zhang, L. Wu, Z. Hu, L. Shi, S. Zhou. Quadruple perovskite ruthenate as a highly efficient catalyst for acidic water oxidation, *Nat. Commun.* 10 (2019) 3809.
[59] X. Xu, Y. Chen, W. Zhou, Z. Zhu, C. Su, M. Liu, Z. Shao. A perovskite electrocatalyst for efficient hydrogen evolution reaction, *Adv. Mater.* 28 (2016) 6442–6448.
[60] V. Mani, S. Anantharaj, S. Mishra, N. Kalaiselvi, S. Kundu. Iron hydroxyphosphate and Sn-incorporated iron hydroxyphosphate: Efficient and stable electrocatalysts for oxygen evolution reaction, *Catalysis Science & Technology* 7 (2017) 5092−5104.

[61] Y. Jiao, Y. Zheng, M. Jaroniec, S. Z. Qia. Design of electrocatalysts for oxygen- and hydrogen-involving energy conversion reactions, *Chem. Soc. Rev.* 44 (2015) 2060–2086.

[62] Q. Tang, D. Jiang. Mechanism of hydrogen evolution reaction on 1T-MoS_2 from first principles, *ACS Catalysis* 6 (2016) 4953–4961.

[63] M. I. Jamesh. Recent progress on earth abundant hydrogen evolution reaction and oxygen evolution reaction bifunctional electrocatalyst for overall water splitting in alkaline media, *Journal of Power Sources* 333 (2016) 213–236.

[64] R. L. Doyle, M. E. G. Lyons. An electrochemical impedance study of the oxygen evolution reaction at hydrous iron oxide in base, *Physical Chemistry-Chemical Physics* 15 (2013) 5224–5237.

[65] M. Gong, H. Dai. A mini review of NiFe-based materials as highly active oxygen evolution reaction electrocatalysts, *Nano Research* 8 (2015) 23–39.

[66] B. M. Hunter, H. B. Gray, A. M. Muller. Earth-abundant heterogeneous water oxidation catalysts, *Chemical Reviews* 116 (2016) 14120–14136.

[67] Y. Cheng, S. P. Jiang. Advances in electrocatalysts for oxygen evolution reaction of water electrolysis-from metal oxides to carbon nanotubes, *Progress in Natural Science: Materials International* 25 (2015) 545–553.

[68] B. Sljukic, C. E. Banks and R. G. Compton. An overview of the electrochemical reduction of oxygen at carbon-based modified electrodes, *J. Iran. Chem. Soc.* 2 (2005) 1–25.

[69] M. Risch. Perovskite electrocatalysts for the oxygen reduction reaction in alkaline media, Catalysts 7 (2017) 154–185.

[70] K. A. Stoerzinger, M. Risch, J. Suntivich, W. M. Lu, J. Zhou, M. D. Biegalski, H. M. Christen, Ariando, T. Venkatesan, Y. Shao-Horn. Oxygen electrocatalysis on (001)-oriented manganese perovskite films: Mn valency and charge transfer at the nanoscale, *Energy Environ. Sci.* 6 (2013) 1582–1588.

[71] S. Priyatharshni, M. Divagar, C. Viswanathan, D. Mangalaraj, N. Ponpandianz. Electrochemical simultaneous detection of dopamine, ascorbic acid and uric acid using $LaMnO_3$ nanostructures, *J. Electrochem. Soc.* 163 (2016) B460-B465.

[72] Z. Zhao, H. Wu, H. He, X. Xu, Y. Jin. Self-standing non-noble metal (Ni–Fe) oxidenanotube array anode catalysts with synergistic reactivity for high-performance water oxidation, *J. Mater. Chem. A* 3 (2015) 7179–7186.

[73] A. Baciu, A. Remes, E. Ilinoiu, F. Manea, S. J. Picken and J. Schoonman. Carbon nanotubes composite for environmentally friendly sensing, *Environm. Eng. Manag. J.* 11 (2012) 239–246.

[74] Y. Gorlin, T. F. Jaramillo. A bifunctional nonprecious metal catalyst for oxygen reduction and water oxidation, *J. Am. Chem. Soc.* 132 (2010) 13612–13614.

[75] Y. Matsumoto, E. Sato. Electrocatalytic properties of transition metal oxides for oxygen evolution reaction, *Materials Chemistry and Physics* 14 (1986) 397–426.

[76] I. Concina, Z. H. Ibupoto, A. Vomiero. Semiconducting Metal Oxide Nanostructures for Water Splitting and Photovoltaics, *Advanced Energy Materials* 7 (2017) 1700706.
[77] P. W. Menezes, C. Panda, S. Loos, F. Bunschei-Bruns, C. Walter, M. Schwarze, X. Deng, H. Dau, M. Driess. A structurally versatile nickel phosphite acting as a robust bifunctional electrocatalyst for overall water splitting, *Energy Environ. Sci.* 11 (2018) 1287−1298.
[78] S. H. Ahn, I. Choi, H.-Y. Park, S. J. Hwang, S. J. Yoo, E. Cho, H.-J. Kim, D. Henkensmeier, S. W. Nam, S.-K. Kim, J. H. Jang. Effect of morphology of electrodeposited Ni catalysts on the behavior of bubbles generated during the oxygen evolution reaction in alkaline water electrolysis, *Chem. Commun.* 49 (2013) 9323−9325.
[79] Y. Zhao, S. Chen, B. Sun, D. Su, X. Huang, H. Liu, Y. Yan, K. Sun, G. Wang. Graphene-Co_3O_4 nanocomposite as electrocatalyst with high performance for oxygen evolution reaction. *Sci. Rep.* 5 (2015) 7629.
[80] C. Liu, H. Ma, M. Yuan, Z. Yu, J. Li, K. Shi, Z. Liang, Y. Yang, T. Zhu, G. Sun, H. Li, S. Ma. $(NiFe)S_2$ nanoparticles grown on graphene as an efficient electrocatalyst for oxygen evolution reaction, *Electrochim. Acta* 286 (2018) 195−204.
[81] Y. H. Li, P. F. Liu, L. F. Pan, H. F. Wang, Z. Z. Yang, L. R. Zheng, P. Hu, H. J. Zhao, L. Gu, H. G. Yang. Local atomic structure modulations activate metal oxide as electrocatalyst for hydrogen evolution in acidic water, *Nat. Commun.* 6 (2015) 8064.
[82] S. Chen, S.-Z.Qiao. Hierarchically porous nitrogen-doped graphene-$NiCo_2O_4$ hybrid paper as an advanced electrocatalytic water splitting material, *ACS Nano* 7 (2013) 10190−10196.
[83] Q. Li, S. Tang, Z. Tang, Q. Zhang, W. Yang. Microwave-assisted synthesis of $FeCoS_2$/XC-72 for oxygen evolution reaction, Solid State Sci. 96 (2019) 105968.
[84] I. Sebarchievici, B. O. Taranu, M. Birdeanu, S. F. Rus, E. Fagadar-Cosma. Electrocatalytic behaviour and application of manganese porphyrin/gold nanoparticle- surface modified glassy carbon electrodes, *Appl. Surf. Sci.* 390 (2016) 131−140.
[85] I. Sebarchievici, A. Lascu, G. Fagadar-Cosma, A. Palade, I. Fringu, M. Birdeanu, B. Taranu, E. Fagadar-Cosma. Optical and electrochemical-mediated detection of ascorbic acid using manganese porphyrin and its gold hybrids, C.R. Chim. 21 (2018) 327-338.
[86] B. R. Kozub, N. V. Rees, R. G. Compton. Electrochemical determination of nitrite at a bare glassy carbon electrode; why chemically modify electrodes?, *Sens. Actuators B Chem. 143* (2010) 539–546.
[87] M. Lucero, M. Riquelme, G. Ramirez, M. C. Goya, A. G. Orive, A. H. Creus, M. C. Arevalo, M. J. Aguirre. A new modified electrode with a copolymer of aniline/Fe(III)-tetrakis(para-aminophenyl)porphyrin: Test of its electrocatalytic activity toward the reduction of molecular oxygen and oxidation of ascorbic acid and sulfite ion, *Int. J. Electrochem. Sci.* 7 (2012) 234−250.

[88] M. C. Gallo, B. M. Pires, K. C. F. Toledo, S. A. V. Jannuzzi, E. G. R. Arruda, A. L. B.Formiga, J. A. Bonacin. The use of modified electrodes by hybrid systems gold nanoparticles/Mn-porphyrin in electrochemical detection of cysteine, *Synth. Met.* 198 (2014) 335–339.
[89] B. O. Taranu, E. Fagadar-Cosma, I. Popa, N. Plesu, I. Taranu. Adsorbed functionalized porphyrins on polyaniline modified platinum electrodes. Comparative electrochemical properties, *Dig. J. Nanomater. Bios.* 9 (2014) 667−679.

CHAPTER 5

POLYCRYSTALLINE CREDNERITE MATERIALS

MARIA POIENAR, IOSIF MALAESCU, CATALIN N. MARIN, PAULINA VLAZAN

Crednerite ($CuMnO_2$) belongs to the ABO_2 class of materials, but with a different crystalline structure due to the Jahn-Teller effect of Mn^{3+}, which lowers the symmetry from the space group R-3m to C2/m. The main difference between the two structures lies in the topology of the B triangular lattice (the Mn^{3+} triangular sublattice is not regular, but anisotropic) and the interplane coupling. In this chapter, the relationships between the structure and properties in these compounds are addressed by corroborating our own results with those presented in the scientific literature. In addition, the impact of the crystallite size and morphology of the obtained $CuMnO_2$ materials on the observed physicochemical properties and the substitution on the electric properties will be discussed.

5.1. Introduction

Numerous research studies have been devoted to the study of ABO_2-type materials (A^+ = Cu or Pd; B^{3+} = transition metal) with delafossite structures because of their various physical properties. Such properties include p-type semiconductivity [1], thermoelectric conversion [2], negative thermal expansion [3], multiferroicity [4-6], DSC cells [7, 8], and catalysis [9, 10]. The delafossite materials present a 2D anisotropic crystal structure characterized by alternating layers along the *c* axis of A cations (A = Cu or Pd), without oxygen, and layers of BO_6 octahedra (B = transition metal) [11]. However, a different structure is encountered for $CuMnO_2$ with space group C2/m due the Jahn-Teller distortion of Mn^{3+} ions in the structure. This determines a deformed triangular network that is different to the isotropic triangular network of ABO_2 delafossites (space group R-3m). The crednerite structure of $CuMnO_2$ displays a large difference between the Mn-O distances in the MnO_6 octahedron with two long apical distances ($Mn-O_{ap}$) corresponding to the direction of the occupied d_{z2} orbitals and four short equatorial distances ($Mn-O_{eq}$). This is therefore different to the regular BO_6 octahedra seen in delafossite structures [12].

$CuMnO_2$ materials possess multiple properties, such as oxygen storage capacity [13]. They are possible candidates for catalysis, as an active catalyst for H_2 in visible light [14, 15], and the oxygen intercalation/deintercalation is evidence in these mterials [16].

On the other hand, crednerite materials have also attracted the interest of the scientific community due to their 'exotic' magnetic and structural properties. $CuMnO_2$ has been shown to support the chemical composition range, i.e. $Cu_{1-x}Mn_xO_2$ [17, 18], where changes are induced in the magnetic structure determined by the modification of the Cu/Mn ratio: the magnetic coupling in the (*a, b*) plane remains the same, but the exchange coupling between the layers changes from antiferromagnetic to ferromagnetic [17, 19, 20]. Furthermore, has been shown that at $T_N = 65$ K in $CuMnO_2$, long range 3D magnetic ordering occurs simultaneously with a lowering of monoclinic symmetry to triclinic symmetry one [20, 21].

It is well known that the size of particles in a material can influence applications, such as those relating to photoelectrochemistry, for example. In this regard, for $Cu_{1+x}Mn_{1-x}O_2$ (x = 0-0.08) polycrystalline materials the value of the ferromagnetic component decreases with increasing x [17, 20]. For nanocrystalline materials $CuMn_{1-x}M_xO_2$ (M = Mg and Al; x = 0.00- 0.04), the magnetic frustration is removed and substitution leads to a small increase in magnetization values at low temperatures, contrary to the fact that the particle size decreases [22].

$CuMnO_2$ is usually prepared by a solid state reaction in sealed silica tubes [14, 17] with Ar [16] or N_2 [13], by low-temperature hydrothermal synthesis [23, 24]. The flux method is used to obtain $Cu_{1.04}Mn_{0.96}O_2$ crystals [18]. More recently, Poienar et al. used the ultrasound-assisted co-precipitation route to obtain $CuMnO_2$ semiconductor nanomaterials [25] that display p-type semiconductivity behavior at room temperature.

A new synthesis technique is presented in this chapter for preparing crednerite $CuMnO_2$ compounds by high-pressure and high-temperature hydrothermal synthesis. Such crednerite materials are obtained from the precursor materials of Cu_2O and Mn_2O_3 and using NaOH as a mineraliser by the one-step hydrothermal method. A number of factors affect the formation processes and product morphologies, including the starting precursors, the temperature, the reaction time, and the pressure. These aspects have all been studied by us and are presented in the following.

In the literature, the synthesis of ABO_2 materials by high-temperature and high-pressure hydrothermal synthesis has been reported for some materials. $AgFeO_2$ was obtained starting from Ag_2O and α-Fe_2O_3 using a basic solution at 400 °C and a pressure of approximatively 2,800 bar [26]. 3R-$CuAlO_2$ has been synthesised from 0.1 mol/Kg $Cu(NO_3)_2$, 0.1 mol/Kg $Al(NO_3)_3$, 1.0 mol/Kg HCOOH, and 0.2 mol/kg NaOH at 400 °C and 30 MPa (300 bar) for 30 min [27]. Recently, Xiong et al. [28] reported the synthesis of $CuAlO_2$ starting from CuCl si $NaAlO_2$ at 320-400 °C and from Cu_2O and Al_2O_3 at 340-400 °C, respectively [28]. The stability diagram for $CuAlO_2$ above and below 250 °C is presented in [29] where 3R-$CuAlO_2$ was obtained at 350 °C and 1,000 bar. Thus, to our knowledge, no data has yet been presented in the literature for the synthesis of $CuMnO_2$.

In the second part of the results and discussion section, the electrical properties of the substitued materials are presented and discussed.

5.2. Experimental Section

In the process of synthesis, precursors with the desired stoichiometry were carefully weighed and then, in the case of soluble metals and salts, were dissolved separately in distilled water before being brought together through continuous magnetic stirring. The NaOH solution was added to this solution, which determines the precipitation. In the case of oxides as starting precursors, these were grounded and transferred directly into the autoclave with the NaOH solution being added afterwards. In the first step, Kennedy diagrams were used to determine the fill factors of the autoclaves and thus the pressure at the given temperature. The autoclaves were made in our laboratory and are schematized in fig. 5-1. For some of the synthesis performed in this study, the pressure was displayed on a sensor (self-powered Plug-on Display, P = 0-1,600 bar), which allowed us to ascertain the exact pressure at which the reaction occurred, and an industrial high-pressure transmitter was also used in our set-up. After the synthesis was finished, the obtained precipitates were filtered, washed with distilled water, and dried at 90 °C for 2 h.

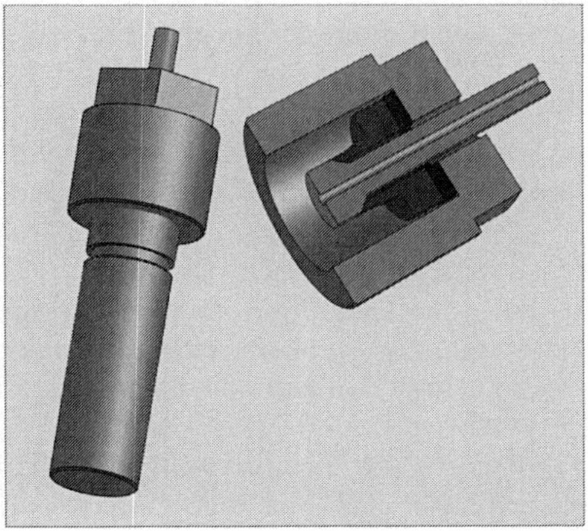

Fig. 5-1. Schematic illustration of the autoclave used for the high-pressure and high-temperature synthesis of $CuMnO_2$.

The RT X-ray powder diffraction (XRD) data were collected using a PANalytical X'pert Pro diffractometer (Cu Kα, $10^o \leq 2\theta \leq 70^o$). Rietveld refinements of the powder diffractograms were performed with the FullProf program from the FullProf suite [30]. The scanning electron microscope images were obtained using the FEI Inspect S microscope. The Fourier transform infrared (FT-IR) spectra were recorded with a JASCO 430 in the range 4,000-400 cm^{-1} using the KBr pellet technique.

In order to study the optical properties of the crednerite-type materials, the powders were analyzed using a Perkin Elmer Lambda 35 UV-Vis spectrometer in atmospheric conditions at wavelengths between 300 and 800 nm.

5.3. Results and Discussion

5.3.1. Synthesis and structural characterizations of $CuMnO_2$ materials obtained by hydrothermal synthesis at high pressure and temperature

Developing synthesis methods for materials with controlled nano or micro particle sizes is essential for achieving their potential. For the class of materials in this study, conventional methods of synthesis for $CuBO_2$ materials include solid-state reactions, the hydrothermal method, the sol-gel method, and co-precipitation, etc. However, solid-state synthesis results in large and agglomerated particles and in solution, the best conditions have to be found in order to control the oxidation states of Cu so that it has +1.

To clarify the effect of reaction temperature, pressure, and the precursors used, a series of experiments were carried out to determine the optimal conditions for crednerite synthesis at high temperature and pressure.

In the first approach, hydrothermal synthesis at high temperature and pressure was performed starting with nitrates $Cu(NO_3)_2.3H_2O$ and $Mn(NO_3)_2.4H_2O$. It should also be mentioned that crednerite materials can be obtained by hydrothermal synthesis at low pressure and temperature [23].

Table 5-1 presents the experimental conditions along with the results for the obtained materials. Thus, we can see that at a temperature of 300 °C and pressure of 50 bar, and at 350 °C and 300 bar, crednerite compounds are obtained as the principal phase, but some impurities are also detected, i.e. Mn_3O_4, CuO. The $CuMnO_2$ phase is not obtained by increasing the temperature to 400 °C at a pressure of 200 bar. The morphology of the sample obtained at $T = 300$ °C and 50 bar is shown in fig. 5-2 where a platelet-like shape is observed for the crednerite particles.

Table 5-1. Experimental conditions for hydrothermal synthesis at high temperature and pressure starting from nitrates $Cu(NO_3)_2.3H_2O$ and $Mn(NO_3)_2.4H_2O$.

Temperature	Pressure	Thermal treatment	Phases
300 °C	50 bar		**$CuMnO_2$**+Mn_3O_4
350 °C	300 bar	12 h, descent in 10 h.	**$CuMnO_2$**+CuO
400 °C	200 bar		CuO+$CuMn_2O_4$

Fig. 5-2. SEM images for the samples obtained by high temperature and pressure starting from nitrates $Cu(NO_3)_2.3H_2O$ and $Mn(NO_3)_2.4H_2O$, $T = 300$ °C and P = 50 bar.

The desired materials were synthesized starting from Cu_2O and Mn_2O_3 at temperatures of 300, 350, and 400 °C, and pressure – between 50 and 300 bar. As can be seen in tables 5-2 and 5-3, the parameters were changed in the experiments to study their effect on the phase formation. The SEM images for some of these samples are given in fig. 5-3.

Table 5-2. Experimental conditions for hydrothermal synthesis at high temperature and pressure starting from Cu_2O and Mn_2O_3, time = 12 h.

Temperature	Pressure	Thermal treatment	Phases
300 °C	50 bar	12 h, descent in 10 h.	**$CuMnO_2$**+Mn_3O_4
300 °C	300 bar		**$CuMnO_2$**+Mn_3O_4
350 °C	400 bar		$CuMnO_2$+MnO
350 °C	1000 bar		$CuMnO_2$+Mn_3O_4
400 °C	300 bar		**$CuMnO_2$**

Table 5-3. Experimental conditions for hydrothermal synthesis at high temperature and pressure starting from Cu_2O and Mn_2O_3, time = 72 h.

Temperature	Pressure	Thermal treatment	Phases
350 °C	300 bar	72 h, descent in 10 h	Mn_3O_4
350 °C	1000 bar		Mn_3O_4+ **$CuMnO_{2\,(s)}$**
350 °C	1000 bar	72 h, descent by inertia	**$CuMnO_2$**+MnO

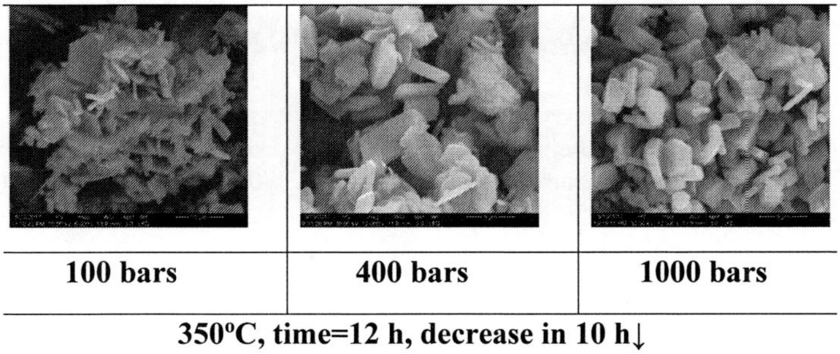

Fig. 5-3. SEM images for the samples obtained by high temperature and pressure starting from Cu_2O and Mn_2O_3.

Table 5-4. Experimental conditions for hydrothermal synthesis at high temperature and pressure starting from already synthesized crednerite

Temperature	Pressure	Thermal treatment	Phases
	370 bar	12 h, descent in 10 h.	**$CuMnO_2$**+Mn_3O_4
	580 bar	72 h, descent in 10 h.	**$CuMnO_2$**
350 acicular	350 bar	72 h, descent in 10 h.	Mn_3O_4
	500 bar	144 h, descent in 10 h.	**$CuMnO_2$**+Mn_3O_4
	500 bar	12 h, descent in 144 h.	**$CuMnO_2$**

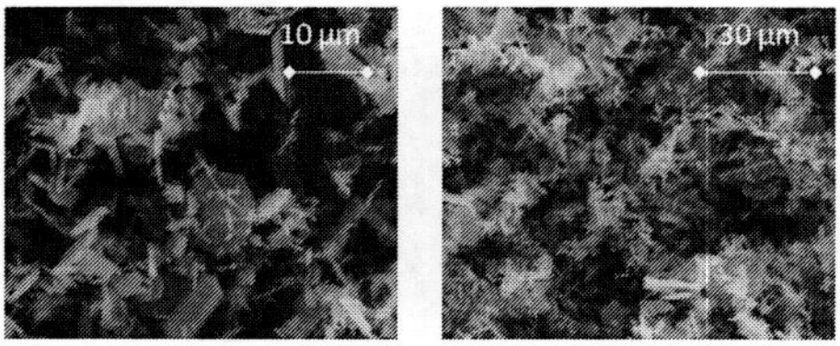

a)

b)

Fig. 5-4. a) Crednerite phase obtained by the hydrothermal method at low pressure and temperature and b) crednerite samples after hydrothermal treatment at high temperature and pressure $T = 350$ °C (up: P = 370 bar; down: 580 bar).

It is reported in the literature that single crystals of crednerite with an acicular shape and being of several millimeters in length, have been obtained by the flux method [18]. Several experiments have been performed in this study using the hydrothermal method at high temperature and pressure to obtain crednerite single crystals. One approach has been to test the effect on the crednerite phase already obtained by the hydrothermal method at low temperature and pressure [22] (fig. 5-4 a)) and determine if single crystals could be obtained. The experimental conditions are presented in table 5-4, but unfortunately only micron-sized crystallites can be grown as observed in fig. 5-4 b)).

The Rietveld refinement for the X-ray diffraction pattern registered for the sample obtained at 400 °C and 200 bar is shown in fig. 5-5. The sample appears well crystalized with peaks that are sharp and very well defined. The unit cell parameters were determined as follows: a = 5.5725(1) Å; b = 2.8745(6) Å; c = 5.8731(1) Å; β = 104.074(1)°; unit-cell volume = 91.252(3) Å^3; and space group = C2/m (according to the literature) [12].

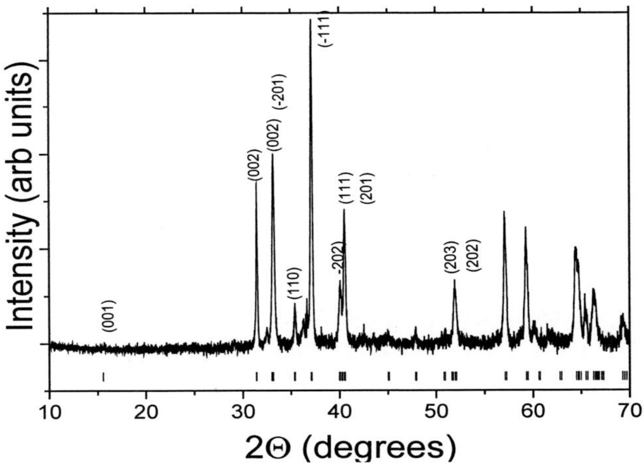

Fig. 5-5. The Rietveld refinement for the crednerite sample obtained at 400 °C and 200 bar.

The FT-IR spectra of the $CuMnO_2$ compound display a band of low intensity around 427 cm^{-1}, a wide band with its maximum located at approximately 536 cm^{-1}, and a narrow band at 724 cm^{-1} corresponding to the vibrations of wagging, bending, and stretching in MnO_n polyhedra (fig. 5-6) [31]. This shows that the sample was well crystallized. In the case of the sample obtained by the hydrothermal method at low pressure and temperature in [22], the bands were observed at: 425, ~510 and ~734 cm^{-1}, similar to those observed for the sample from this study. The large bands observed around 1,000 cm^{-1} could be due to some residual material from the solution.

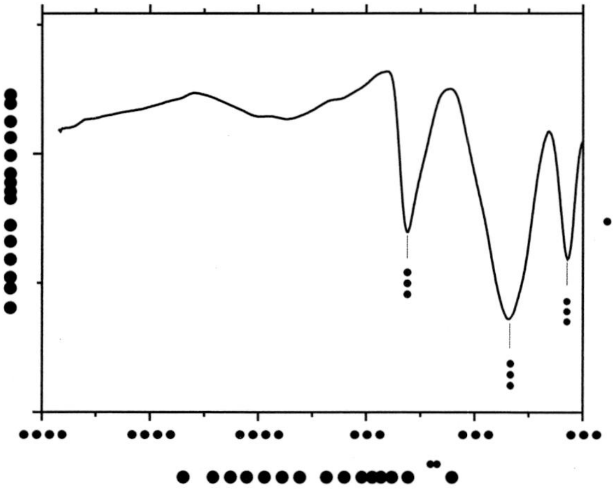

Fig. 5-6. The FT-IR spectra for the crednerite sample obtained at 400 °C and 200 bar.

5.3.2. The AC conductivity of $CuMn_{1-x}AlxO_2$ (x = 0-0.06) type crednerite materials prepared by the hydrothermal method at low temperature and pressure

Crednerite $CuMn_{1-x}M_xO_2$ (M = Al; x = 0-0.06) samples were prepared according to the modified hydrothermal method described in the literature [22, 32]. The conditions for the synthesis of these materials were: T = 100 °C for 24 h. in a Teflon-lined stainless steel autoclave (filled to 80 %) and using the following precursors dissolved in distilled water: $Cu(NO_3)_2.3H_2O$, (1-x) $Mn(NO_3)_2.4H_2O$, and x $Al(NO_3).9H_2O$, (x = 0-0.06).

Figure 5-7 presents the structural characterisations of the obtained powders according to Raman spectroscopy. In the literature, two Raman active modes have been observed at ~300 and 700 cm^{-1} and are considered to be characteristic of ABO_2 type compounds [33, 34]. These findings are in agreement with the X-Ray analysis, which shows that the phases are well crystallized and no impurity peaks were detected.

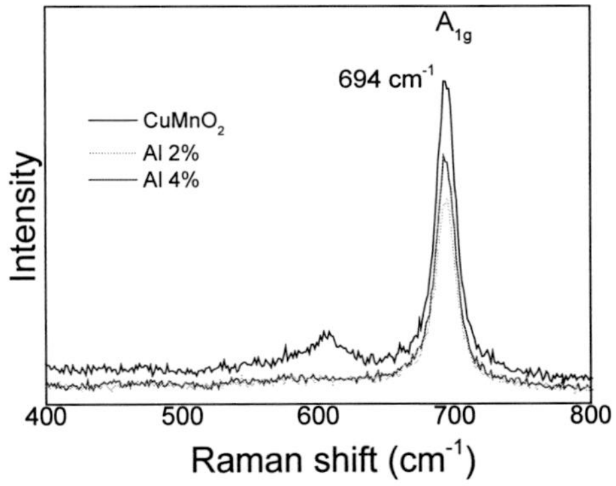

Fig. 5-7. Raman spectra for the $CuMn_{1-x}Al_xO_2$ (x = 0, 0.02, and 0.04) samples.

Using UV-VIS spectroscopy we were able to determine, for the first time, the band gap value for the pristine and Al-substituted crednerite samples. For this, we used the Tauc equation: (αhν) 1/n = A (hν-E_g), where h is the Plank constant; ν is the frequency of the incident light; α is the absorption coefficient; A is the proportionality constant; and E_g is the energy band gap defined as the difference between the level with the lowest energy in the conduction band and the level with the highest energy in the valence band [35].

Figure 5-8 presents the optical absorption spectra in the wavelength range of 300-800 nm for $CuMn_{1-x}M_xO_2$ (M = Al; x = 0-0.06). Using a linear fit, the value for the energy band gap is calculated as 3.48 eV for $CuMnO_2$, 3.52 eV for 2 % Al, 4 % Al, and 3.55 eV for 6 % Al.

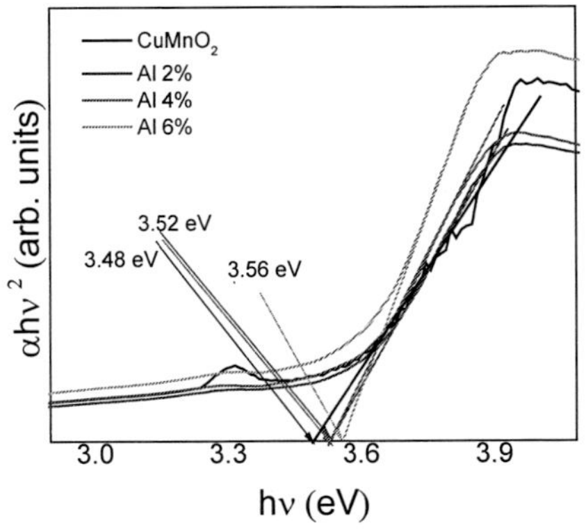

Fig. 5-8. Calculation of E_g for $CuMn_{1-x}Al_xO_2$ (x = 0.0-0.06) obtained at 100 °C, for 24 h.

Change in the values for Eg is generally determined by three effects: the quantum effect of size; the surface effect; and/or the electron-phonon coupling. In the case of semiconductors, a redshift for E_g could be due to the quantum effect of decreasing particle size and the surface effect, while a blushift for E_G is induced by electron-phonon coupling [36]. In our case, the different particle size induced by substitution may explain the slight increase in the Eg value for 2 %, 4 %, and 6 % Al crednerite-substituted samples. Poienar et al. showed that the Mg and Al substitution induces a decrease in particle size in crednerite samples [22].

Measurements of complex impedance were performed for the $CuMn_{1-x}Al_xO_2$ (x = 0, 0.02, 0.04, and 0.06) samples, using the technique of complex impedance spectroscopy (CIS) described in Chapter 2. The experimental setup is presented in the following.

The investigated sample was placed in a cylindrical glass tube with length *L* and cross-section *A*. It was in contact with two electrodes, which allowed the connection of an RLC-meter [37, 38]. At room temperature *T*, the real component *Z′* and imaginary component *Z″* of the complex impedance of the sample were measured. Figure 5-9 shows the frequency dependence of the real (*Z′*) and imaginary (*Z″*) components, of the complex impedance of the samples in the frequency range 20 Hz-2 MHz) and at different concentrations x of Al ions substituted in the $CuMnO_2$ crednerite material (x = 0, 0.02, 0.04, and 0.06).

As can be seen in fig. 5-9, at low frequencies up to 1 kHz, the real component Z′ increased with increasing substituted Al ion concentration x. This behaviour shows that Al substitution in $CuMnO_2$ material leads to a decrease in electrical conductivity for all x concentrations. The imaginary component Z″ displays a maximum at frequency $f_{max,(Z)}$, which moves to lower values when concentration x increases. The appearance of this maximum shows the existence of an electrical relaxation process [39, 40] due to the presence of charge carriers in the samples.

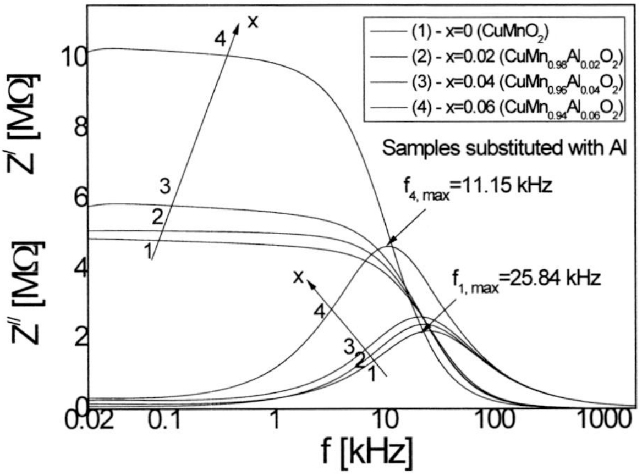

Fig. 5-9. The frequency dependence of the real (Z′) and imaginary (Z″) components of the complex impedance of $CuMn_{1-x}Al_xO_2$ crednerite samples at x = 0, x = 0.02, x = 0.04, and x = 0.06.

The amplitude of *Z″* increases with an increase in x concentration and this behaviour is in good agreement with a small decrease of Cu^{+}-Cu^{+} ionic distance (lattice parameter *a*) [41] with the substitution of Mn^{3+} by Al^{3+}. This can observed by Rietveld refinement of XRD data [41, 22], which leads to a decrease in DC conductivity corresponding to low frequencies.

Using the Debye equation $2\pi f_{\max}\tau_Z = 1$ [42] and the experimental values of $f_{max,(Z)}$, in fig. 5-9 we have computed the relaxation times τ_Z corresponding to each x concentration of Al substituted ions in $CuMnO_2$ material. The obtained values are listed in table 5-5.

In table 5-5, we can observe an increase of the relaxation time τ_Z by increasing x concentration of Al ions from 0 to 0.06. This increase in τ_Z can be correlated with a decrease in the lattice volume with increasing concentration (x) of Al ions, observed via X-ray diffraction analysis [22] and considering that the substitution of Mn with Al ions affects only of the Mn sublattice [43].

The relaxation mechanism of electrical conduction in powders of $CuMn_{1-x}Al_xO_2$ type was also investigated by analyzing the frequency dependence of the real (M′) and imaginary (M″) components of the electric modulus *M.* Knowing, the measured values Z′ and Z″ of the complex impedance and the geometry of the sample (length $l = 8$ mm and transversal sectional area A = 11.33 mm²), the values M′ and M″ of the electric modulus M can be computed with the relations:

$$M' = \left(\frac{\omega\varepsilon_0 A}{l}\right)\cdot Z'' \quad (5.1)$$

$$M'' = \left(\frac{\omega\varepsilon_0 A}{l}\right)\cdot Z' \quad (5.2)$$

where ω is the angular frequency and ε_0 is the permitivity of the free space.

Table 5-5. Experimental and calculated values of $CuMn_{1-x}Al_xO_2$ crednerite samples with x = 0, x = 0.02, x = 0.04, and x = 0.06.

Samples	$CuMnO_2$	$CuMn_{0.98}Al_{0.02}O_2$	$CuMn_{0.94}Al_{0.06}O_2$	$CuMn_{0.96}Al_{0.04}O_2$
$\tau_Z\,[\mu s]$	6.16	6.68	14.46	7.19
$\tau_M\,[\mu s]$	5.53	6.21	12.57	6.70
$\sigma_{DC}\,[S/m]$	$1.447\cdot10^{-4}$	$1.385\cdot10^{-4}$	$0.694\cdot10^{-4}$	$1.219\cdot10^{-4}$
n	0.38	0.41	0.50	0.42
$A[S/m]$	$9.48\cdot10^{-7}$	$5.29\cdot10^{-7}$	$1.10\cdot10^{-7}$	$4.73\cdot10^{-7}$
$\omega_c[rad/s]$	$5.58\cdot10^{5}$	$7.89\cdot10^{5}$	$3.98\cdot10^{5}$	$5.50\cdot10^{5}$
$a[m]$ [6]	$5.554\cdot10^{-10}$	$5.543\cdot10^{-10}$	$5.536\cdot10^{-10}$	$5.537\cdot10^{-10}$
$N_C[m^{-3}]$	$5.12\cdot10^{24}$	$3.48\cdot10^{24}$	$3.47\cdot10^{24}$	$4.40\cdot10^{24}$
$\mu[m^2/V\cdot s]$	$0.176\cdot10^{-9}$	$0.248\cdot10^{-9}$	$0.125\cdot10^{-9}$	$0.173\cdot10^{-9}$

Figure 5-10 shows the frequency dependence of the real (M') and imaginary (M'') components of the electric modulus of the samples in the frequency range 20 Hz-2 MHz and at different concentrations x of Al ions substituted in the $CuMnO_2$ crednerite material (x = 0, 0.02, 0.04, and 0.06).

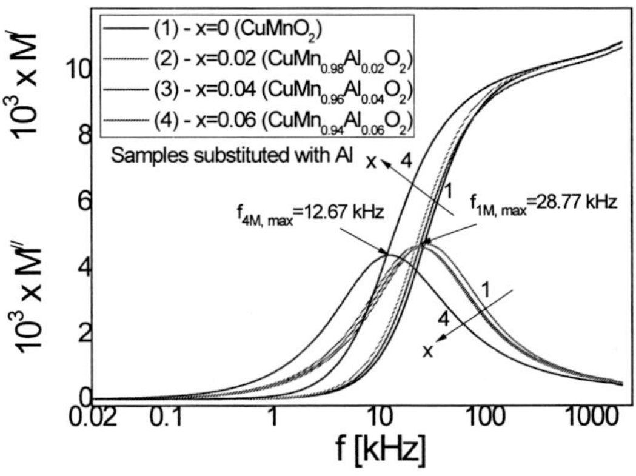

Fig. 5-10. The frequency dependence of the real (M′) and imaginary (M″) components of the electric modulus of the $CuMn_{1-x}Al_xO_2$ crednerite samples with x = 0, x = 0.02, x = 0.04, and x = 0.06.

As can be seen in fig. 5-10, in the low frequency range of 20 Hz and 0.2 kHz, *M′* is very small, approaching zero for all concentrations x of Al ions. This shows reduced polarization effects at the interface electrode/sample [44, 45]. By increasing the frequency, the component *M′* increases continually and tends towards the value M_∞ at each concentration x. Also, in fig. 5-10 we can observe that with increasing frequency, the magnitude of *M′* increases slightly with increasing concentration. This can be attributed to a concentration-dependent relaxation process in the sample. The imaginary component *M″* shows a peak centered in the dispersion region of *M′* for each concentration x. Each maximum of *M″* corresponds to the relaxation frequency $f_{max,(M)}$, which moves from 28.77 kHz to 12.67 kHz when the Al ion concentration increases from x = 0 to x = 0.06. This shows a shift towards frequencies greater than the frequencies corresponding to the maximum $f_{max,(Z)}$ of the imaginary component *Z″* of the complex impedance [46], as can be seen in fig. 3-11.

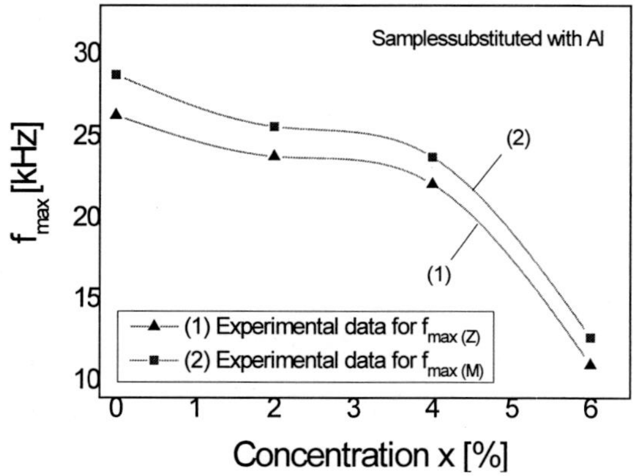

Fig. 5-11. The concentration dependence of the $f_{max,(Z)}$ and $f_{max,(M)}$ frequencies, corresponding to the imaginary components Z″ and M″, respectively.

Using the experimental values of $f_{max,(M)}$ and the Debye equation, we computed the relaxation times τ_M corresponding to each concentration x of Al substituted ions in $CuMnO_2$ material. The obtained values are listed in table 5-5. As a result, we have for the first time characterized the DC and AC conductivity of the crednerite samples. Based on the experimental values Z′ and Z″ of the complex impedance (fig. 5-9), the total conductivity σ of the samples was determined, with the equation:

$$\sigma = \frac{Z'}{|Z|^2} \cdot \frac{l}{A} \tag{5.3}$$

where $|Z| = \sqrt{Z'^2 + Z''^2}$ is the complex impedance modulus of the sample.

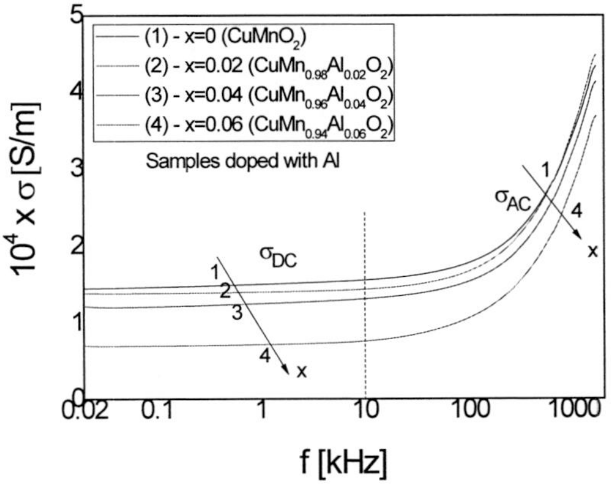

Fig. 5-12. The frequency dependence of the conductivity σ at different concentrations x of Al ions substituted in the $CuMnO_2$ material.

The frequency dependence of the conductivity σ in the frequency range 20 Hz-2 MHz and at different concentrations x of Al ions substituted in the $CuMnO_2$ crednerite material is shown in fig. 5-12.

As can be observed in fig. 5-12, at a constant concentration x of the Al ions substituted in $CuMnO_2$ material, the conductivity spectrum, $\sigma(f)$ has of two regions: 1) a region in which σ does not vary with the frequency corresponding to DC-conductivity (σ_{DC}); and 2) a dispersion region where σ rapidly increases with frequency corresponding to AC conductivity (σ_{AC}). This behaviour is in accordance with Jonscher's universal law [47]:

$$\sigma(\omega) = \sigma_{DC} + \sigma_{AC} \tag{5.4}$$

In the first region (fig. 5-12), σ_{DC} depends only on the x concentration of Al ions substituted in $CuMnO_2$ material, remaining approximately constant with the frequency up to about 10 kHz. The values obtained for σ_{DC} are listed in table 5. We can see that varying the concentration x from x = 0 to x = 0.06 leads to a decrease in conductivity from $1.447 \cdot 10^{-4}$ *S/m* to $0.694 \cdot 10^{-4}$ *S/m* for $CuMn_{1-x}Al_xO_2$ samples. The values obtained for DC-conductivity are similar to those obtained by other authors for $CuFe_{1-x}Cr_xO_2$ ($0<x<1$) delafosite samples [41], or $Cu_{1+x}Mn_{1-x}O_2$ crednerite samples (with x = 0-0.10) [48, 49, 50] where the substitution determines the increase in electrical resistivity.

In the dispersion region, AC-conductivity (σ_{AC}) depends on the frequency and correlates to the dielectric relaxation processes determined by localized electric charge carriers from the sample [47]. It is given by:

$$\sigma_{AC} = A\omega^n \tag{5.5}$$

where *A* is a pre-exponential factor and *n* is an exponent dependent on both frequency and temperature ($0<n<1$) [51]. As a result, taking into account equations (5.4) and (5.5), Jonscher's universal law can be written in another form:

$$\sigma(\omega) = \sigma_{DC} + A\omega^n = \sigma_{DC}\left[1+\left(\frac{\omega}{\omega_c}\right)^n\right] \tag{5.6}$$

where the pre-exponential factor $A = \omega_c^{-n}\sigma_{DC}$ and ω_c is the crossover frequency [41]. In agreement with the dynamic Jonscher universal response [47], ω_c represents the frequency of the transition from the DC regime with low frequencies to the dispersive regime of high frequencies, also called the hopping frequency ω_h.

Logarithmic transformation of the equation (3.5) results in a linear dependence between $ln\sigma_{AC}$ and $ln\omega$:

$$\ln\sigma_{AC} = \ln A + n\ln\omega \tag{5.7}$$

This is presented in fig. 5-13 for the investigated samples.

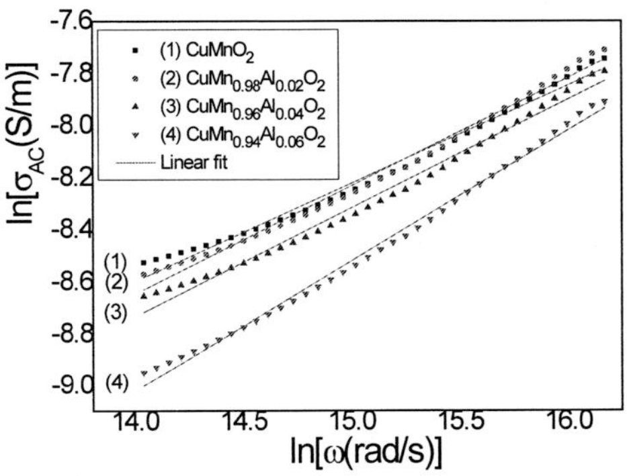

Fig. 5-13. The $ln\sigma_{ac}(ln\omega)$ dependence at different x concentrations for $CuMn_{1-x}Al_xO_2$ samples.

By fitting the experimental dependence $ln(\sigma_{ac})(ln(\omega))$ from fig. 5-13 with a straight line, we determined the parameter *A* and the exponent *n* corresponding to each concentration x of substituted Al ions. The obtained values are shown in table 5-5. Knowing the values of *A* and *n*, as well as σ_{DC} conductivity, we determined the hopping frequency ω_h corresponding to each concentration x of Al ions substituted in the $CuMnO_2$ sample. The values obtained for ω_h are shown in table5-5.

Table 5-5 shows a slight increase in the exponent *n* from 0.38 for $CuMnO_2$ (x = 0) to 0.50 for $CuMn_{1-x}Al_xO_2$ (x 0.06). The low variation of *n* shows reduced dispersion of σ_{AC} with frequency and increasing x concentration of Al ions (see fig. 5-12). This behaviour in the investigated samples may be correlated with a possible decrease in particle size as a result of the substitution of Mn^{3+} by Al^{3+}, which can lead to a contraction of the unit cell, and a decrease in the lattice parameter *a*, as Poienar et al. recently observed [22].

It is known that for the hopping transport of charge carriers [52], the relationship between the conductivity σ_{DC} and the hopping frequency ω_h [41, 52] is given by:

$$\sigma_{DC} = \left(\frac{N_C \cdot e^2 \cdot r_h^2}{12\pi \cdot k_B T} \right) \cdot \omega_h \tag{3.8}$$

where r_h is the hopping distance, i.e. the Cu^+-Cu^+ ionic distance *a* [53]; k_B is the Boltzmann constant; *e* is the charge of the electron; and N_C is the number density of the effective charge carrier [22, 53]. Considering the values of the lattice parameter *a* for the samples $CuMn_{1-x}Al_xO_2$ (with x = 0, 0.02, 0.04, and 0.06), obtained from Rietveld refinements using XRD data [22] (see table 5-5), and assuming that $a = r_h$ [22, 53], the number density of the effective charge carrier N_C can be evaluated using relation (3.8). The obtained values are shown in table 5-5.

The charge carrier mobility μ can be determined by the σ_{DC} and N_C values using the following relation:

$$\mu = \frac{\sigma_{DC}}{eN_C} \tag{3.9}$$

and the obtained values are shown in table 5-5. The x concentration Al ion-dependence of the charge carrier mobility μ is presented in fig. 5-14.

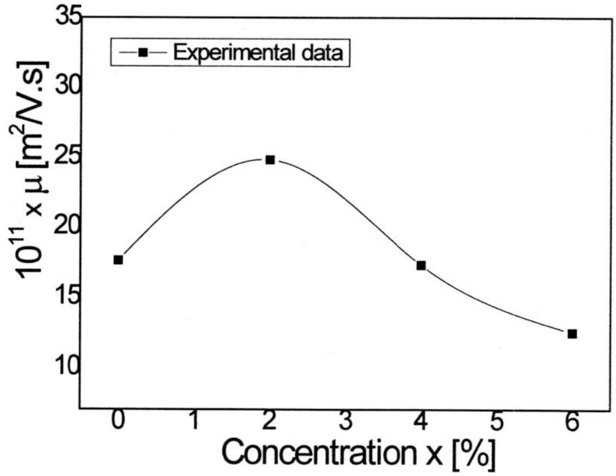

Fig. 5-14. The concentration (x) dependence of charge carrier mobility (μ) for the $CuMn_{1-x}Al_xO_2$ samples.

In agreement with Jonscher's universal power law [47], the frequency dependence of AC conductivity at different concentrations of the Al ions substituted in the $CuMnO_2$ material (fig. 5-12) corresponds to the short-range hopping of charge carriers between sites separated by energy barriers of variable height [54]. The relatively low values obtained for the exponent *n* (*n* < *0.5*) of the investigated samples show the repeated hopping of the charge carriers between the nearest neighbor states. In order to correlate the conduction mechanism of AC conductivity with the temperature dependence of exponent *n*, at different concentrations x of the substituent, a number of theoretical models have been proposed [54, 55, 56]. Therefore, in this paper we used the correlated barrier hopping (CBH) theoretical model, as proposed by Pike [57], for a single electron hop in the $CuMn_{1-x}Al_xO_2$ crednerite material at a constant room temperature *T*, but different concentrations x of Al ions substituted in the $CuMnO_2$ sample. According to the CBH model, the exponent *n*, in a first approximation, can be written as:

$$n = 1 - \frac{6kT}{W_m} \qquad (5.10)$$

where W_m represents the maximum barrier height (equal to the band gap energy) [58]. Using equation (5.10), we determined the x concentration dependence of the energy barrier W_m, as shown in fig. 5-15.

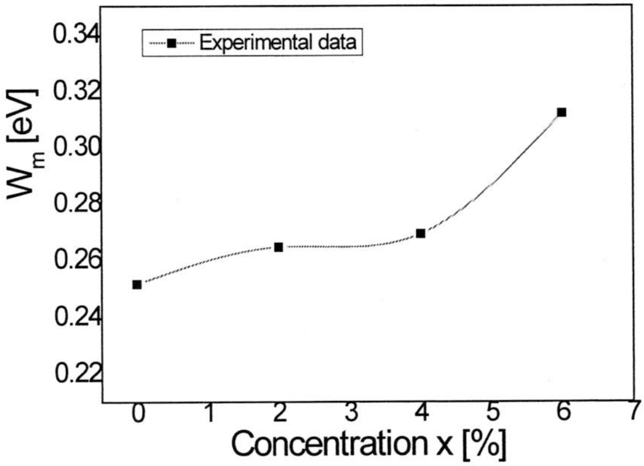

Fig. 5-15. The concentration (x) dependence of the band gap energy (W_m) for the $CuMn_{1-x}Al_xO_2$ samples.

5.4. Conclusion

In this chapter, we presented the high-temperature and high-pressure hydrothermal synthesis of $CuMnO_2$ compounds for the first time. One of the main conclusions is that the crystal size of crednerite materials is not influenced by an increase in pressure or/and temperature.

The structure and phase composition of the samples were studied by Raman and FT-IR spectroscopy, which revealed that the samples obtained at 400 °C and ~200 bar crystallized within the crednerite structure with space group C2/m.

Measurements of the AC and DC conductivity were performed for crednerite Al ion-substituted materials for the first time in this study and their mechanism has been explained.

Acknowledgments

This work was supported by a grant from the Romanian National Authority for Scientific Research and Innovation, CNCS–UEFISCDI, project number PN-II-RU-TE-2014-4-2179. The authors thank D. Ursu for the X-ray diffraction measurements and A. Dabici for the UV-Vis measurements.

References

[1] H. Kawazoe, M. Yasukawa, H. Hyodo, M. Kurita, H. Yanagi and H. Hosono. *Nature* 389 (1997) 939.

[2] K. Koumoto, H. Koduka and W.S. Seo. *J. Mater. Chem.* 11 (2001) 251.

[3] J. Li, A. Yokochi, T. G. Amos, and A. W. Sleight. *J. Chem. Mater.* 14 (2002) 2602–2606.

[4] K. Kimura, H. Nakamura, K. Ohgushi and T. Kimura. *Phys. Rev. B* 78 (2008) 140401.

[5] M. Poienar, F. Damay, C. Martin, V. Hardy, A. Maignan and G. André. *Phys. Rev. B* 79 (2009) 014412.

[6] T. Kimura, J. C. Lashley, A. P. Ramirez. *Phys. Rev. B* 73 (2006) 220401 (R).

[7] M. Yu, G. Natu, Z. Ji, and Y. Wu. *J. Phys. Chem. Lett.* 3 (9) (2012) 1074–1078.

[8] M. Yu, T. I. Draskovic and Y. Wu. *J. Phys. Chem. Chem. Phys.* 16 (2014) 5026–5033.

[9] C. Mondelli, A. P. Amrute, T. Schmidt and Javier Pérez-Ramírez. *Chem. Commun.* 47 (2011) 7173.

[10] K. Gurunathan, Jin-Ook Baeg, Sang, Mi Lee, E. Subramanian, Sang-Jin Moon, Ki-Jeong Kong. *Catalysis Communications*, Volume 9, Issue 3, 1 March 2008, 395.

[11] W. Soller, A. J. Thompson. *Phys. Rev.* 47 (1935) 644.

[12] I. D. Kondrashev. *Soviet Physics Crystallography* 3 (1959) 703.

[13] S. Kato, R. Fujimaki, M. Ogasawara, T. Wakabayashi, Y. Nakahara and S. Nakata. *Appl. Catalysis B: Environmental* 89 (2009) 183.

[14] Bessekhoud, M. Trari, J. P. Doumerc. *Int. J. Hydrogen Energy* 28 (2003) 43–48.

[15] M. Poienar, R. Banica, P. Sfirloaga, C. Ianasi, C.V. Mihali, P. Vlazan. *Ceramics International* 44 (2018) 6157–6161.

[16] B. Bellal, B. Hadjarab, N. Benreguia, Y. Bessekhouad, M. Trari. *J. Appl. Electrochem.* 41 (2011) 867–872.

[17] M. Trari, J. Töpfer, P. Dordor, J.-C. Grenier, M. Pouchard and J.P. Doumerc. *J. Solid State Chem.* 178, 2751 (2005).

[18] J. Töpfer, M. Trari, P. Gravereau, J. P. Chaminade and J.P. Doumerc. *Zeit. Kristallogr.* 210, 184 (1995).

[19] F. Damay, M. Poienar, C. Martin, A. Maignan, J. Rodriguez-Carvajal, G. André and J. P. Doumerc. *Phys. Rev. B* 80, 094410 (2009).

[20] M. Poienar, C. Vecchini, G. André, A. Daoud-Aladine, I. Margiolaki, A. Maignan, A. Lappas, L. Chapon, M. Hervieu, F. Damay, C. Martin. *Chemistry of Materials* 23, 85-94 (2011).

[21] C. Vecchini, M. Poienar, F. Damay, O. Adamopoulos, A. Daoud-Aladine, A. Lappas, J. M. Perez-Mato, L. C. Chapon, and C. Martin. *Phys. Rev. B* 82 (2010) 094404.

[22] M. Poienar, P. Sfirloaga, C. Martin, D. Ursu, P. Vlazan. *Journal of Materials Science* 53 (2018) 2389–2395.

[23] M. Poienar, C. Martin, O. I. Lebedev, A. Maignan. *Solid State Sciences*, Volume 80, June 2018, 39–45.
[24] D. Xiong, Q. Zhang, Z. Du, S. Kumar Verma, H. Lia and X. Zhao. *New J. Chem.* 40 (2016) 6498--6504.
[25] M. Poienar, A. Lungu, P. Sfirloaga, M. Lungu, C. V. Mihali and P. Vlazan. *Chemical Papers* 73 (6) (2019) 1541–1546.
[26] W. J. Croft, N. C. Tombs and R. E. England. *Acta Crystt.* 17 (1964) 313.
[27] T. Sato, K. Sue, H. Tsumatori, M. Suzuki, S. Tanaka, A. Kawai-Nakamura, K. Saitoh, K. Aida, T. Hiaki. *J. of Supercritical Fluids* 46 (2008) 173–177.
[28] D. Xiong, X. Zeng, W. Zeng, H. Wang, X. Zhao, W. Chen. *Inorganic Chemistry* 53 (2014) 4106–4116.
[29] M. Miclau, N. Miclau, R. Banica and D. Ursu. *Materials Today: Proceedings* 4 (2017) 6975–6981.
[30] J. Rodriguez-Carvajal. *Physica B* 192 (1993) 55–69.
[31] C. M. Julien, M. Massot and C. Poinsignon. *Spectrochim. Acta Part A* 60 (2004) 689–700.
[32] Q. Zhang, D. Xiong, H. Li, D. Xia, H. Tao, X. Zhao, J Mater Sci: Mater Electron 26 (2015) 10159–10163.
[33] O. Aktas, K. D. Truong, T. Otani, G. Balakrishnan, M. J. Clouter, T. Kimura and G. Quirion. *J. Phys.: Condens. Matter* 24 (2012) 036003.
[34] Manoj K. Singh, S. Dussan, Ganpat L. Sharma, and Ram S. Katiyara. *Journal of Applied Physics* 104, 113503 (2008).
[35] J. Tauc, R. Grigorovici, A. Vancu. *Phys. Status Solidi B* 15 (1966) 627−637.
[36] K. Oshiro, K. Akai, M. Matsuura. *Phys. Rev. B* 59, 10850 (1999).
[37] A. Lungu, I. Malaescu, C. N. Marin, P. Vlazan, P. Sfirloaga. *Phys. B* 462 (2015) 80–85.
[38] ASTM D150-98: Standard test methods for AC loss characteristics and permittivity (dielectric constant) of solid electrical insulation.
[39] M. Nadeem, M. J. Akhtar, A. Y. Khan. *Solid State Commun.* 134 (2005) 431-436.
[40] I. Malaescu, A. Lungu, C.N. Marin, P. Sfirloaga, P. Vlazan, S. Brindusoiu, M. Poienar. *Ceramics International* 44 (2018) 11610–11616.
[41] M. Lalanne, P. Demont, A. Barnabé. *Journal of Physics D: Applied Physics* 44 (18) (2011) 185401.
[42] P. Debye. *Polar Molecules*, Chemical Catalog Company, New York (1929).
[43] R. D. Shannon, D. B. Rogers, C. T. Prewitt. *Inorg. Chem.* 10 (1971) 713–718.
[44] M. H. Harun, E. Saion, A. Kassim, E. Mahmud, M. Y. Hussain, I. S. Mustafa. *J. Adv. Sci. Arts* 1 (1) (2009) 9–16.
[45] T. Horn, S. Deutschlander, H. Lowen, G. Maret, and P. Keim, Phys. Rev. E 88 (2013) 062305.
[46] M. Wubbenhorst, Jan van Turnhout. *J. Non-Cryst. Solids* 305 (2002) 40–49.
[47] A. K. Jonscher. *Universal Relaxation Law* (1st edn.) Chelsea Dielectrics Press: London (1996).
[48] Y. Bessekhouad, Y. Gabes, A. Bouguelia, M. Trari. *J. Mater. Sci.* 42 (15) (2007) 6469–6476.
[49] H. Chen, D. Hsu. *Applied Surface Science* 290 (2014) 161–166.

[50] K. Shukla, P. Shahi, S. Gopal, A. Kumar, A. K. Ghosh, R. Singh, N. Sharma, A. Das, A. K. Sinha, A. Joshi, A. Nigam and S. Chatterjee. *RSC Adv.* 5 (2015) 83504–83511.
[51] K. Funke. *Prog. Solid State Chem.* 22 (1993) 111–195.
[52] N. F. Mott, E. A. Davis. *Electronic Processes in Non-crystalline Materials*, Clarendon: Oxford (1979).
[53] M. Lalanne, A. Barnabé, F. Mathieu, Ph. Tailhades. *Inorg. Chem.* 48 (2009) 6065.
[54] J. C. Dyre, T. B. Schroder. *Rev. Mod. Phys.* 72 (2000) 873.
[55] T. Okuda, N. Jufuku, S. Hidaka, N. Terada. *Phys. Rev. B* 72 (2005) 144403.
[56] R. Nagarajan, A.D. Draeseke, A.W. Sleight, J. Tate, J. Appl. Phys. 89 (2001) 8022.
[57] G. E. Pike. *Phys. Rev. B* 6 (1972) 1572–1580.
[58] S. R. Elliott. *Philos. Mag. B* 36 (6) (1977) 1291–1304.

CHAPTER 6

APPLICATIONS

CATALIN N. MARIN, MIHAIL LUNGU, MADALIN BUNOIU, IOSIF MALAESCU

This chapter focuses on the applications of oxide (perovskite, spinel, and crednerite) materials related to renewable energy, photocatalysis, or gas sensors, for example. The relationships between their structures and properties will be emphasised along with the influence of the synthesis methods or substitution/doping on the observed electric and magnetic properties. Following these reports, new perspectives on the technological and biomedical applications of these oxide materials can be envisaged.

6.1. Introduction

Oxygen forms stable chemical bonds with nearly all chemical elements giving oxide compounds. Due to the chemical composition of our planet, hydrolysis and oxidation are natural chemical processes in the formation of oxide materials, which is why much of the earth's crust consists of oxide compounds. The abundant presence of natural oxides and the environmental conditions that favour the formation of oxides are, perhaps, the reasons why so many oxide materials are found to have uses in various applications.

Iron oxides were among the first oxide materials used by humans in the form of pigments in cave paintings and for painting ancient pottery, for example. Commonly available colours included yellow, red, and black, along with their combinations.

Magnetite (Fe_3O_4) is an oxide material that was used in manufacturing one of the greatest inventions of mankind—the magnetic compass. Developed during the Han Dynasty (ca. 206 BC) [1], the magnetic compass with a lodestone was used in navigation much later, starting in the 11th century AD [2]. It made a major contribution to the development of shipping and commercial exchange between different regions of the world.

6.2. Electrical Engineering Applications

With the evolution of ideas about electromagnetism, oxide materials began to be used in many technical applications. The discovery of the law of electromagnetic induction [3] and its use in the production of electricity created the need to develop permanent magnets and magnetic cores. For magnetic cores, iron was the first option. Later on, with the increase in power and working frequency, it was understood that magnetic cores with low electrical conductivity were needed to reduce eddy currents. In this regard, **soft ferrites** became the first-choice materials for magnetic cores used in the electronics industry to make efficient high-frequency inductors, transformers, and antennas, as well as in various microwave components. Manganese-zinc ferrite and nickel-zinc ferrite are commonly used soft ferrites.

Manganese-zinc ferrite (with the formula $Mn_xZn_{(1-x)}Fe_2O_4$) is characterized by high values of the real part of relative magnetic permeability μ' in the order of 10,000-12,000 and small values of the imaginary part of relative magnetic permeability μ'' of the order of 10^{-5}-10^{-6}, over a wide frequency range from zero to tens of kHz [4].

Nickel-zinc ferrite (with the formula $Ni_xZn_{(1-x)}Fe_2O_4$) is widely used in the high frequency range due to its high electrical resistivity and ferromagnetic resonance frequency in comparison to manganese-zinc ferrite [4]. Depending on the Zn content and the heat treatment temperature, Ni-Zn-ferrites have values of relative magnetic permeability between the tens and hundreds. The substitution of nickel with zinc leads to increased permeability and decreased frequency of use [4].

It has been found that **lithium** and **lithium-zinc ferrites** can be used at higher frequencies than nickel-zinc ferrites. Up to 100 MHz they show low losses and relative magnetic permeability values of between 10 and 100 [5].

In contrast to soft ferrites, which have a small coercive field and low remnant magnetization, **hard ferrites** generate a hysteresis curve characterized by the high value of the coercive field and strong remnant magnetization. Magnetically hard oxide materials, which are used in the manufacturing of permanent magnets, fall into two categories: cobalt ferrite-based materials and hexagonal-ferrite based materials (with barium, strontium, and lead ferrite) [4].

Cobalt ferrite magnets were first obtained by Kato and Takei in 1933. Such magnets are made from a mixture of magnetite and cobalt ferrite by pressing and sintering at temperatures of 1,000-1,100 °C for 1-2 hours, followed by slow cooling. Cooling from 300 °C to room temperature is done in an intense magnetic field. In this way, magnets with coercive fields of 400-600 Oe and residual induction of 3,000-5,000 Gs are obtained [4]. Due to the fact that cobalt is a limited material, the cost price of these magnets is high. In addition, such magnets are fragile to mechanical shocks and vibrations.

The most common permanent magnets using hard magnetic ferrites are made with **strontium ferrite**, $SrFe_{12}O_{19}$ ($SrO \cdot 6Fe_2O_3$), and **barium ferrite**, $BaFe_{12}O_{19}$ ($BaO \cdot 6Fe_2O_3$). Strontium and barium ferrites are both hexagonal ferrites with the structure of magnetoplumbite [7]. They are widely used in a number of applications due to the abundance of the raw materials and their low price. The remanent induction of barium ferrite magnets is 380 mT [4] and for strontium ferrite this can reach 400 mT [8]. Even though they have less residual induction than rare earth magnets, ferrite magnets are used in electric motors, recording media, magneto-optic media, loudspeakers, magnetic stripe cards, and many other applications.

6.2.1. Electromagnetic absorbers

The development of telecommunications in the radio and microwave range has resulted in an ocean of electromagnetic waves with consequences for possible interferences between equipment and devices, as well as possibly on the health of human beings and on the environment. Starting from this factual situation, research in the field of microwave electromagnetic absorbers has gained a distinctive and special place in the area of electromagnetism.

There are different types of electromagnetic absorbers, some are resonant absorbers, such as the Salisbury screen, the Jaumann absorber, the Dallenbach layer, crossed-grating absorbers, and circuit analogue (CA) absorbers, while others are so-called broadband absorbers. The design of broadband electromagnetic absorbers is based on theoretical models that describe the frequency behaviour of the complex magnetic permeability and complex dielectric permittivity of the materials and on the measurements of electromagnetic wave propagation parameters in different materials subjected to various external conditions.

The propagation of electromagnetic waves through a medium is described by the wave equation, written either in the form for the electric field component (eq. (6.1)), or in the form for the magnetic field component (eq. (6.2)) [10].

$$\Delta\vec{E} = \gamma^2\vec{E} \tag{6.1}$$

$$\Delta\vec{H} = \gamma^2\vec{H} \tag{6.2}$$

In these equations, $\vec{E}$ is the intensity of the electric field of the electromagnetic wave; $\vec{H}$ is the intensity of the magnetic field of the electromagnetic wave; and γ represents the electromagnetic wave propagation constant. $\gamma = \alpha + j\beta$ is a complex quantity with the real part α constituting the attenuation constant and its imaginary part β constituting as the phase constant. The electromagnetic wave propagation constant and its real and imaginary components can be expressed in terms of complex relative magnetic permeability as $\mu_r = \mu_r' - j\mu_r''$ and in terms of complex relative dielectric permittivity as $\varepsilon_r = \varepsilon_r' - j\varepsilon_r''$ of the propagation medium [11]:

$$\gamma = j2\pi f\sqrt{\varepsilon_0\mu_0}\cdot\sqrt{(\mu_r' - j\mu_r'')(\varepsilon_r' - j\varepsilon_r'')} \tag{6.3}$$

$$\alpha = \pi f\sqrt{\mu_0\varepsilon_0}\sqrt{2\left[\sqrt{(\mu_r'^2 + \mu''^2_r)(\varepsilon'^2_r + \varepsilon''^2_r)} - (\mu_r'\varepsilon_r' - \mu_r''\varepsilon_r'')\right]} \tag{6.4}$$

$$\beta = \pi f\sqrt{\mu_0\varepsilon_0}\sqrt{2\left[\sqrt{(\mu'^2_r + \mu''^2_r)(\varepsilon'^2_r + \varepsilon''^2_r)} + (\mu_r'\varepsilon_r' - \mu_r''\varepsilon_r'')\right]} \tag{6.5}$$

In eqs. (6.3)-(6.5) $j = \sqrt{-1}$, f is the frequency of the electromagnetic wave; μ_0 and ε_0 are the free space permeability and permittivity; ε_r' and ε_r'' are the real and imaginary components of the complex relative dielectric permittivity of the propagation medium; and μ_r' and μ_r'' are the real and imaginary components of the complex relative magnetic permeability of the propagation medium.

In analyzing the properties of materials designed as electromagnetic absorbers, four parameters are usually evaluated: i) the reflection coefficient at the air-material interface R; ii) the overall reflection coefficient R_w of the material with the thickness w deposited on a total reflective support; iii) the quarter wavelength in material $\lambda_{1/4}(f, H)$; and sometimes, iv) the shielding effectiveness *SE* of the material.

The reflection coefficient at the air-material interface R is defined by equation (6.6) for the normal incidence of the electromagnetic wave,with the assumption that it comes from the medium "1" and is reflected onto the surface of the medium "2" [11]. The coefficient R shows the ability of materials to reflect electromagnetic waves.

$$R = \left| \frac{\sqrt{\frac{\mu_{r2}\varepsilon_{r1}}{\mu_{r1}\varepsilon_{r2}}} - 1}{\sqrt{\frac{\mu_{r2}\varepsilon_{r1}}{\mu_{r1}\varepsilon_{r2}}} + 1} \right| \tag{6.6}$$

The absorption coefficient α shows the ability of a material to attenuate electromagnetic waves, but the overall reflection coefficient with normal incidence (defined by equation (6.7)) shows the ability of a material to attenuate the wave while passing through the absorber before and after total reflection on a reflective support.

$$Rw = R + (1 - R)^2 \exp(-2\alpha w) \tag{6.7}$$

For practical use of the overall absorption coefficient *Rw*, the possible dimensional resonance must be taken into account and, therefore, the quarter wavelength in the material $\lambda_{1/4}(f, H)$ must always be evaluated. $\lambda_{1/4}(f, H)$ is correlated with the phase constant β of the wave in the material, This is given by the relation (6.8). Once $\lambda_{1/4}(f, H)$ is evaluated, *Rw* is relevant for a material whose thickness is smaller than $\lambda_{1/4}$.

$$\lambda_{1/4} = \frac{\pi}{2\beta} \tag{6.8}$$

According to the standard ASTM D:4935-10 [12], the shielding effectiveness of a material *SE* is a quantity defined as the ratio of the power of an electromagnetic wave P1, received in the presence of the material used as a screen, and the power of the electromagnetic wave P2, received in the absence of such material, for the same incident power. The shielding effectiveness is usually expressed in decibels:

$$SE = 10\log\left(\frac{P_1}{P_2}\right) \tag{6.9}$$

If evaluation of the electromagnetic wave is done in electric field strength, then the electromagnetic shielding effectiveness *SE* is expressed according to the relation (6.8).

$$SE = 20\log\left(\frac{|\vec{E}_1|}{|\vec{E}_2|}\right) \tag{6.10}$$

Here, $\vec{E}_1$ and $\vec{E}_2$ represent the intensity of the electric field for the received wave in the presence of the material used as a screen and in the absence of the material for the same incident power of the electromagnetic wave.

6.2.2. Solar cells

Metal oxides have been applied in photovoltaic technologies for many years. Their remarkable properties and simple manufacturing methods give oxides a unique place in next-generation photovoltaics (NGPV).

A highly relevant class of oxides for photovoltaic conversion is constituted by transition metal oxides due to their catalytic activity and semiconductor properties. Oxides of transition metals are compounds composed of oxygen atoms bonded to a metal with a partially occupied sub-shell. In practice, these can be considered elements of groups 3 to 12 in the periodic table. The basic structural units of the oxides of the transition elements are octahedral MO_6 and tetrahedral MO_4. The d orbitals of the metal overlap with the 2p orbitals of the ligand forming molecular orbitals [13].

Complex oxides of transition metal materials are distinguished by their wide range of magnetic and electronic properties, such as ferromagnetism, ferroelectricity, antiferroelectricity, and high temperature superconductivity. These properties often come from the strongly correlated electrons between the d or f orbitals. Antiferoelectric photovoltaics [14] is a growing field of research and contains the class of materials with strong photovoltaic effects [15]. The photoelectric voltage in this type of solar cell is not limited by the forbidden band, resulting in photo-tensions of over 10 kV for $LiNbO_3$ crystals [16]. The photovoltaic field can have values that are a few orders of magnitude larger than those usually given by a p-n junction. It has been shown that such values can even reach orders of the range MV/cm (an order of magnitude larger than the critical electric field of the most common semiconductors, such as silicon) [17].

Due to the high band gap energy range, there are many oxides that are potentially suitable for use as "light grabbers". The most well-studied of these is silicon. Other possible substances have been discovered that have historically received much more attention, but only a few metal oxides have been seriously investigated as photon absorbers. Among these metal oxides, copper oxide (CuO) remains the most popular material. The stable binary oxide phases of copper oxides are Cu_2O (copper oxide), CuO (cupric oxide), and Cu_4O_3, and the prohibited CuO_x band can be adjusted to cover the entire range between 1.4 and 2.2 eV [18]. Emerging oxides that are being investigated as light absorbers include cobalt oxides (Co-O) and iron oxides (Fe-O) [19].

Binary oxides can easily be obtained from low-cost materials that can be processed in solution. The evolving research behind this new method of synthesis at low temperatures has led to the successful manufacture of non-crystalline (amorphous) oxides with good electrical properties and now applied in various photovoltaic and optoelectronic devices. These oxides are excellent materials to be applied in printed and electronic solar cells. As complex oxides, they possess important properties including ferroelectricity, magnetism, and pyroelectricity among others, which, together with ordinary semiconductor materials (organic, inorganic, or hybrid) can be used to develop solar cells with new characteristics and properties. Many new features are yet to be discovered and this will only be possible if the synthesis of these complex oxides can be performed at low temperatures.

A useful characteristic of semiconductor oxides is that they can confer stability on next-generation solar cells, especially the most advanced form: perovskite solar cells. Thousands of hours of device stability have been reported, taking us a step further towards commercializing this technology.

Many properties of semiconductor oxides have been discovered, but there are many more to discover. The development of low-temperature, low-cost processing methods can allow the development of new and exciting properties, stimulating the use of semiconductor oxides in future photovoltaic research.

6.2.3. LEDs and photodetectors

Light emitting diodes (LED) are optoelectronic devices that convert electricity into light energy directly through the electro-optical effect. They are constructed of a junction formed by direct contact between an n-type semiconductor material (in which the majority carriers are negatively charged electrons) and a p-type semiconductor material (in which the majority charge carriers are positively charged voids).

One-dimensional (1D) metal oxide nanostructures are ideal systems for studying a large number of new nanoscale phenomena and investigating the size and dimensionality dependence of nanostructure properties for potential applications. The construction of photodetectors or optical switches based on 1D metal oxide nanostructures with adapted geometries has developed a lot in recent years. The elements of a photodetector with an active 1D nanostructure are configured either as field effect transistors whose properties can be controlled by applying appropriate voltages, or as resistors whose conductors are modified by a load transfer process. The functionalization of the surfaces of the structure offers another way to expand the capabilities of such sensors [20].

6.3. Chemical Engineering Applications

6.3.1. Water purification

Water purification by adsorption processes. A promising direction in the use of nanostructured oxide nanomaterial and functionalized materials is that of water purification. Scientific progress and industrialization have been both beneficial and problematic for the environment and for human populations. Currently, the biggest problems are those related to water, including depletion of groundwater; contamination of freshwater resources and increased demand for freshwater (due to rapid population growth and urbanization). A sustainable solution to these problems is to treat contaminated water. Contaminated water can contain a wide variety of chemical species from domestic or industrial wastewater that can have harmful effects on the environment or can lead to disease in humans and animals, such as metal ions, dyes, pharmaceuticals, pesticides, etc. [21].

As such, the development of efficient methods of water purification is a very important and topical area of research. Among the methods that can be used for this purpose we may mention: precipitation; ion exchange; the use of membranes; electrochemical methods;

coagulation-flocculation; phyto and bioremediation [22]; photocatalytic processes; advanced oxidation [23]; and adsorption.

Adsorption is a widely used technique for advanced wastewater treatment due to its advantages such as simplicity, high efficiency, and the availability of a wide range of adsorbents. This technique is especially suitable for removing pollutants that cannot be easily biodegraded from water. Current research in the field of water treatment is seeking to develop new materials with advanced adsorbent properties.

Nanostructured metal oxides, such as those formed with iron (Iron(III) oxide-hydroxide FeOOH, hematite α-Fe_2O_3, magnetite γ-Fe_2O_3, and magnetite Fe_3O_4), manganese, aluminium, zinc, titanium, and cerium are all considered promising materials, especially for use as adsorbents for the removal of heavy metals, oxyanions (especially arsenic), dyes, and other organic pollutants from aqueous systems [24-30, 22].

Adsorbent iron-based materials have been studied extensively, especially as adsorbents for arsenic species, due to the affinity between these elements. Given the abundance of arsenic in the environment, reducing its concentration in drinking water is a difficult, topical, and critical issue in many countries around the world. Solving this problem has been the goal of a large number of researchers who have developed a number of materials for the adsorption of arsenic species and iron oxide or hydroxide materials have proven to be very effective.

When adsorbent materials are used to remove pollutants from the water, one of the most difficult problems that arises is the efficient separation of solid particles at the end of the process. Especially in the case of non-biodegradable adsorbents, their separation and recovery is essential to avoid secondary pollution. Of the new-generation materials developed recently, magnetic nanoadsorbents are the most promising. Such materials have a large specific surface area, adjustable morphology, and, above all, have the advantage that they can be easily separated from the aqueous solution by use of a magnetic field. In recent years, spinel ferrites have become the first choice for water treatment applications because they have nanometric dimensions, along with magnetic properties and a high surface/volume ratio [31-34].

In the literature, there are a large number of studies dedicated to the use of magnetic nanoadsorbents for the retention of water-polluting species. Below, we list a few of them:

- Heavy metal ions for Cr (VI) adsorption: magnetite, $MnFe_2O_4$[36], $MgFe_2O_4$, $ZnFe_2O4$, $CuFe_2O_4$, $NiFe_2O_4$, $CoFe_2O_4$ [35].

- Dyes. To retain Congo red: cobalt ferrite doped with gadolinium [37], zinc ferrite [38], cobalt ferrite, nickel, copper [22], $Fe_{3-x}La_xO_4$ ferrite [38] $MnFe_2O_4$ [36], maghemite [39]; $Co_xNi_{1-x}Fe_2O_4$ [40], $Ni_{0.5}Zn_{0.5}Fe_2O4$ [41], $ZnFe_2O_4$ [42]; MFe_2O_4 (M = Mn, Fe, Co, Ni) [43]. For the adsorption of acid red 88: $ZnFe_2O_4$ [44]. For the adsorption of methyl blue: $Ni_{0,5}Zn_{0,5}Fe_2O_4$ [45] and $MnFe_2O_4$ [46].
- Other organic pollutants: MFe_2O_4 (M = Fe^{2+}, Mn^{2+}, Co^{2+}, Zn^{2+}) for the adsorption of some tetracyclines [47].
- Arsenic species: $MnFe_2O_4$ and $CoFe_2O_4$ [48] magnetite [49] CuFe2O4 [50].

Water purification by advanced oxidation processes. For the removal of non-biodegradable pollutants or those that degrade with difficulty from water by primary and secondary treatment processes, tertiary treatment is required. This includes advanced oxidation processes, such as chemical oxidation, catalytic oxidation, and thermal oxidation, etc. Advanced oxidation processes are some of the most promising methods for degrading organic pollutants in water. They are based on the generation of reactive species, such as hydroxyl (HO •) and sulphate (SO_4 • 2) radicals, which are strongly oxidant and react non-selectively. Under certain conditions, pollutants can be completely mineralized to harmless products (CO_2, H_2O, and short-chain organic acids, etc.). Radicals are usually generated from H_2O_2, O_3, peroxide monosulphate (PMS), or peroxide disulphate (PDS), using transition metals or ultraviolet/visible radiation (photochemical processes). In particular, the processes that take place in heterogeneous systems have been intensively studied due to the advantages they offer, such as operation at ambient temperatures and normal pressure, and the easy recovery of the catalyst (especially in the case of magnetic materials) for reuse.

6.3.2. Photodegradation

Nanoscale advances in science and engineering suggest that many of the current problems involving water quality could be solved or improved using nanocatalysts as an economical and effective alternative for water purification and decontamination. Nanomaterials are globally available and have a remarkable ability to effectively remove organic and inorganic pollutants from water. However, for the successful commercialization of this water treatment technology, certain inherent shortcomings need to be adequately addressed. These include the aggregation of nanoparticles in the water filtration process and their adverse effects on human health and the environment. In order to overcome these problems and obtain products a sufficient number of desired properties, efforts are being made to design and develop nanostructures.

Research in this area has spawned new and advanced oxidation processes, especially in the form of heterogeneous photocatalysis, converting photon energy into chemical energy. The application of the concept of materials engineering in heterogeneous catalysis is based on the transformations that take place at the surface of the catalyst at the molecular level [51]. Photocatalytic activities have been reported in the last decade for many oxide nanomaterials, such as TiO_2, ZnO, and graphite carbon nitride (g-C_3N_4), etc. [53, 54]. For more than three decades, the most studied of these nanomaterials is TiO_2 due to its excellent photocatalytic activity. Moreover, to ensure the magnetic separation of TiO_2, the incorporation of a Fe_3O_4 core has proven to be very useful [55]. Fe_3O_4, being a photocatalyst in itself, has a super paramagnetic behaviour for particles with a size below 20 nm [56, 57]. This property together with its photocatalytic capacity has been reported to be promising for water and wastewater treatment technologies [58-60].

The use of TiO_2 nanoparticles for the photodegradation of wastewater dyes can be considered a viable alternative due to the fact that the optical, chemical, and electrical properties of photocatalysts are exploited. In practice, however, the efficiency of the photocatalytic process varies quite a lot, depending on the material used, as well as the environmental factors in which the process takes place. The optimization of the photocatalytic process depends on obtaining materials with characteristic properties, such as the crystalline structure, crystallite size, specific surface, and the presence of defects that by correction can give photocatalytic systems high efficiency, short operating times, and the possibility of being used at scale with low production costs.

Similarly, ZnO a broadband semiconductor, and g-C_3N_4, a metal-free photocatalyst in the composition, have recently emerged as effective materials for photocatalytic applications [61-63].

6.3.3. Metal oxides as a cathode material

Researchers need to investigate electrochemical storage devices for electric vehicles and smart grids [64-67]. Such devices are water rechargeable batteries that have become competitive due to their favourable features such as low cost, high safety, and environmental safety [68, 69].

Lithium compounds used as materials for the electrodes of rechargeable aqueous batteries have limited their widespread application due to insufficient resources and high cost [70, 71]. Zinc ion rechargeable aqueous batteries are considered to be promising state-of-the-art batteries. These compounds are based on the zinc compound intercalated as a cathode and zinc as an anode due to the low cost, abundance, and environmental safety of zinc [72-78]. Only a few cathode materials, including spinel $ZnMn_2O_4$ and α-, β-, γ-, δ-, λ-MnO_2, and vanadium oxide have been shown to be intercalating hosts for aqueous rechargeable zinc ion batteries.

$ZnMn_2O_4$ is one of the plausible cathode materials for aqueous rechargeable zinc ion batteries [79-86]. However, studies have shown that the discharge capacity of $ZnMn_2O_4$ is too low without the presence of large amounts of carbon as a conductive agent. The spinel phase is permanently present during the loading/unloading process, while various MnO_2-type polymorphs show low performance in terms of speed, as well as a rapid reduction in capacity. This is due to the poor conductivity of MnO_2, the formation of an irreversible phase of $ZnMn_2O_4$ spinel and the dissolution of manganese during continuous cycling [87-91].

6.4. Medical Applications

Metal oxide nanoparticles are of interest for both basic and applied research due to their biocompatibility, stability, magnetic properties, and simple preparation properties. Thus, they become attractive for applications in various fields such as biomedicine, food, industry, cosmetology, pharmaceuticals, environmental protection, and biotechnology.

There are a large number of publications describing the use of metal oxide nanoparticles in biomedicine. Due to their multifunctional properties, nanoparticles can be successfully applied in diagnosis, drug therapy as nanoelectronic biosensors, and in molecular nanotechnology.

For example, Fe_3O_4 nanoparticles play a crucial role in diagnosis, they can be used as contrast agents in magnetic resonance imaging, for positron emission in tomography, infrared fluorescence, and as biosensors for the detection of biomolecules such as glucose, proteins, urea, and uric acid. The wide implementation potential of Fe_3O_4 nanoparticles is also oriented towards therapy, ranging from cancer treatment, as antibacterial agents, biosensors, in tissue repair, and cell proliferation. With the evolution of recent research in the field of NP Fe_3O_4 nanotechnology, there has been an orientation towards the construction of nanoenzymes that have biological oxidase, peroxidase, catalase, and superoxide dismutase-like activities. Nanosymes are nanomaterials with intrinsic properties similar to biological enzymes, which offer several advantages over natural enzymes, such as high stability, pH conditions, and cost-effectiveness.

ZnO nanoparticles have high applicability in biomedicine and have high potential for the treatment of cancer, diabetes, wound healing, drug delivery, and bioimaging. Due to the inherent toxicity of NP ZnO, they have strong impacts against cancer cells and bacteria by inducing intracellular ROS generation and activating the apoptotic signal, making them a potential candidate as anticancer and antibacterial agents. In addition, NP ZnO are also been known for their use in transporting therapeutic drugs or biomolecules. Moreover, they have the ability to lower blood glucose and increase insulin levels, showing promising potential for treating diabetes and alleviating its complications.

In addition to engineering applications, oxide materials have been found to be useful for various medical applications. Advances in the diagnosis and treatment of various diseases have been and are closely linked to technological advances. Furthermore, radio diagnosis would not have been possible without the discovery of X-rays and image reconstruction techniques in CT imaging; ultrasound scanning would not have been possible without the discovery of the piezoelectric effect; and MRI would not have been possible without the discovery of the phenomenon of nuclear magnetic resonance and the examples can continue. In many of the current techniques of diagnosis and treatment, oxide materials have been involved to a greater or lesser extent.

6.4.1. As a magnetic marker for non-palpable tumours in surgical oncology

Permanent magnets are now being used to mark the spatial position of tumours. Using such screening programs, surgeons can locate and operate on small and non-palpable tumours. Therefore, precise surgical localization has great importance in guiding the surgeon to the correct region. Using a catheter, a small piece of magnet (encapsulated in titanium for biocompatibility) or a small quantity of iron oxide nanoparticles [92] is inserted into the tumour region. With good visibility in all medical imaging systems, this magnetic marker can be implanted for up six months before surgery, thus providing accurate surgical oncology guidance.

6.4.2. Cancer treatment by magnetic hyperthermia of tissues

Metal oxides are increasingly used in medical applications, such as materials for magnetic fluid hyperthermia. The hyperthermia technique involves transporting the oxides to the area of interest and heating them using an external radio frequency or magnetic field until the cells reach temperatures of about 43 °C, destroying cancerous tissue without affecting the surrounding normal tissue [93].

There are a number of biomedical applications using magnetic nanoparticles with ferromagnetic properties of metal oxides, for which Mn, Co, Ni, and Zn are good candidate metals. These nanoparticles are synthesized by various methods to produce controlled shape, desired chemical composition, good dispersion in surfactants, and biocompatibility.

Magnetic nanoparticles are being used in medical applications such as materials for magnetic fluid hyperthermia. The hyperthermia technique consists of delivering nanoparticles to the area of interest and then heating them up using an external radiofrequency or magnetic field until the cells reach 46 °C in order to destroy cancerous tissue without affecting surrounding normal tissue [92].

$MnFe_2O_4$ superparamagnetic nanoparticles were coated with heat-sensitive copolymer and tested for their viability and affinity to carcinoma cells under various conditions. The nanoparticles were loaded with the anticancer drug doxorubicin. Compounds with a diameter of 45 nm have a core-shell structure and a magnetic core of about 18 nm. The effects of magnetic hyperthermia on cell viability and drug release were studied by exposing the cell suspension to a high frequency magnetic field [93].

The drug was released at temperatures above the lower critical temperature of the solution by magnetic heating. It was observed that the lower critical temperature of the thermosensitive copolymer solution is about 39 °C. Below this temperature, the copolymer showed hydrophilic characteristics, increasing its volume. At temperatures above the lower critical temperature of the solution, the copolymer became hydrophobic, eliminating water, drugs, and reducing the volume. The combination of hyperthermia and medication was shown to effectively treat cancer cells.

6.5. Conclusion

Magnetic nanoparticles such as spinel/inverse spinel ferrites have attracted the attention of scientists in the past decade because the methods for obtaining them are simple, cheap, and have a very high degree of reproducibility, with the ability to rigorously control the process parameters. The properties of the end product depends on the morphological and structural parameters: average size of crystallite; ratio of crystalline phases; dopant concentration; and degree of crystallinity. Spinel-type ferrite nanoparticles have applications in the treatment of cancer-hyperthermia, as contrast agents for medical imaging, reversible magnetic nanocatalysts, controlled drug-release, and magnetic quantum dots, etc.

$CoFe_2O_4$ nanoparticles have received special attention because they are used in storage media, memory cores, and high frequency devices due to their unique properties. They have a number of magnetic, electronic, optical, and physical properties, such as high Curie temperature, high magnetocrystalline anisotropy, a magnetostriction coefficient, and high chemical stability. In vitro and in vivo studies have shown that they can be used as biomaterials for magnetic hyperthermia applications [94, 95]. Magnetic hyperthermia is dominated by energy loss through three main heating mechanisms: hysteresis loss, Neel relaxation, and Brown relaxation. These mechanisms are influenced by the average size of the crystallite. Heat dissipation of superparamagnetic nanoparticles is rarely accompanied by loss of hysteresis, while ferromagnetic nanoparticles show a significant loss, improving heating efficiency.

The use of iron oxide nanoparticles in the treatment of cancer by magnetic hyperthermia was first demonstrated by Gilchrist et al. in 1957. In 2004, the first clinical, medical system for the treatment of hyperthermia was developed at Charite Medical University in Berlin. Since then, Magforce has obtained European approval to treat patients with brain tumors using this method [96].

Metal oxide nanoparticles are among the most widely used nanomaterials due to their unique properties. Metal oxides are indispensable tools in modern nanotechnology due to their varied optical properties, with higher ductility than coarse-grained ceramics at high temperatures, cold welding properties, superparamagnetic behavior, unique catalytic, sensitivity, and selective activity.

Nanotechnology is an advancing field that includes engineering and research in developing new materials at the nanoscale level. Several important synthesis strategies and techniques are used in furthering the field of nanotechnology.

References

[1] William Lowrie. *Fundamentals of Geophysics*, Cambridge University Press, London (2007). ISBN:978-0-521-67596-3.

[2] Joseph Needham. *Science and Civilisation in China*, Cambridge University Press, London (1962). ISBN:0-521-05802-3

[3] David Halliday, Kenneth S. Krane Robert Resnick. *Physics* (5th edn.). ISBN: 10:0471320579.

[4] S. Cedighean. Ferite, *Editura Tehnică*, București (1966).

[5] M. S. Ruiz, S. E. Jacobo. Electromagnetic properties of lithium zinc ferrites doped with aluminium, *Physica B* 407 (2012) 3274–3277.

[6] Y. Kato, T. Takei. The Permanent Oxide Magnet and its Characteristics, *The Journal of the Institute of Electrical Engineers of Japan*, 53 (1933) Issue 538, 408–412.

[7] H. Kojima. Fundamental properties of hexagonal ferrites with magnetoplumbite structure, Handb. Ferromagn. *Mater.* 3 (1982) 305–391.

[8] D. Drummer, S. Messingschlager. Material Characterization of Strontium Ferrite Powders for Producing Sintered Magnets by Ceramic Injection Molding (MagnetPIM), *Advances in Materials Science and Engineering*, 2014, article ID 651062. doi:10.1155/2014/651062.

[9] Abdi-Fatah Hersi, Staffan Eriksson, Joakim Ramos, Shahin Abdsaleh, Fredrik Wärnberg, Andreas Karakatsanis. A combined, totally magnetic technique with a magnetic marker for non-palpable tumour localization and superparamagnetic iron oxide nanoparticles for sentinel lymph node detection in breast cancer surgery, *Eur. J. Surg. Oncol.* 45 (2019) 544–549. doi:10.1016/j.ejso.2018.10.064.

[10] R. E. Collin. *Foundation for Microwave Engineering*, McGraw-Hill (1966).

[11] P. C. Fannin, C. N. Marin, I. Malaescu, N. Stefu, P. Vlazan, S. Novaconi, S. Popescu. Effect of the concentration of precursors on the microwave absorbent properties of Zn/Fe oxide nanopowders, *Journal of Nanoparticle Research* 13 (2011) 311–319.

[12] ASTM D4935-99, *Standard Test Method for Measuring the Electromagnetic Shielding Effectiveness of Planar Materials* (1999).

[13] Amador Pérez-Tomás, Alba Mingorance, David Tanenbaum and Mónica Lira-Cantú. Metal Oxides in Photovoltaics: All-Oxide, Ferroic, and Perovskite Solar Cells, *The Future of Semiconductor Oxides in Next-Generation Solar Cells. Metal Oxides* (2018) 267 –356. https://doi.org/10.1016/B978-0-12-811165-9.00008-9.

[14] Y. Yuan et al. Arising applications of ferroelectric materials in photovoltaic devices, *J. Mater. Chem. A* 2 (17) (2014) 6027–6041.

[15] P. J. Sturman, V. M. Fridkin. Photovoltaic and photo-refractive effects in noncentrosymmetric materials, in G.W. Taylor (ed.) *Ferroelectricity and Related Phenomena*, vol. 8, Gordon and Breach Sience Publishers, Philadelphia, PA (1992).

[16] A.M. Glass, D. von der Linde, T. J. Negran. High-voltage bulk photovoltaic effect and the photorefractive process in LiNbO3, *Appl. Phys. Lett.* 25 (4) (1974) 233–235.

[17] A. Perez-Tomas, M. Lira-Cantu, G. Catalan. Above-bandgap photovoltages in antiferroelectrics, *Adv. Mater.* 28 (43) (2016) 9644–9647.

[18] A. Y. Anderson et al. Quantum efficiency and bandgap analysis for combinatorial photovoltaics: sorting activity of Cu–O compounds in all-oxide device libraries, *ACS Comb. Sci.* 16 (2) (2014) 53–65.

[19] S. Ruhle et al. All-oxide photovoltaics, *J. Phys. Chem. Lett.* 3 (24) (2012) 3755–3764.

[20] Zhai, T., Fang, X., Liao, M., Xu, X., Zeng, H., Yoshio, B., and Golberg, D. *A Comprehensive Review of One-Dimensional Metal-Oxide Nanostructure Photodetectors. Sensors*, 9 (8) (2009) 6504–6529. doi:10.3390/s90806504.

[21] Hua M., Zhang S., Pan B., Zhang W., Lu L., Zhang Q. Heavy metal removal from water/wastewater by nanosized metal oxides: A review, *Journal of Hazardous Materials* 211-212 (2012) 317–331.

[22] Ungureanu G., Santos S., Boaventura R., Botelho C. Arsenic and antimony in water and wastewater: Overview of removal techniques with special reference to latest advances in adsorption, *Journal of Environmental Management* 151 (2015) 326–342.

[23] Stoia M., Muntean C., Militaru B. MnFe2O4 nanoparticles as new catalyst for oxidative degradation of phenol by peroxydisulfate, *Journal of Environmental Sciences* 53 (2017) 269–277.

[24] Adegoke H. I., Adekola F. A., Fatoki O. S., Ximba B. J. Sorptive Interaction of Oxyanions with Iron Oxides: A Review, *Polish Journal of Environmental Studies* 22 (1) (2013) 7–24.

[25] Chen L., Xin H., Fang Y., Zhang C., Zhang F., Cao X., Zhang C., Li X. Application of Metal Oxide Heterostructures in Arsenic Removal from Contaminated Water, *Journal of Nanomaterials* (2014a) 793610, 10. http://dx.doi.org/10.1155/2014/793610.

[26] Hashim M. A., Mukhopadhyay S., Sahu J. N., Sengupta B. Remediation technologies for heavy metal contaminated groundwater, Journal of Environmental Management 92 (2011) 2355-2388.

[27] Hua M, Zhang S, Pan B, Zhang W, Lv L, Zhang Q, Heavy metal removal from water/wastewater by nanosized metal oxides: A review, *Journal of Hazardous Materials* 211-212 (2012) 317–331.

[28] Lata S., Samadder S. R. Removal of arsenic from water using nano adsorbents and challenges: A review, *Journal of Environmental Management* 166 (2016) 387–406.

[29] Mohan D., Pittman Jr C. U. Arsenic removal from water/wastewater using adsorbents: A critical review, *Journal of Hazardous Materials* 142 (2007) 1–53.

[30] Mohapatra M., Anand S. Synthesis and applications of nano-structured iron oxides/hydroxides: A review, *International Journal of Engineering, Science and Technology* 2 (8) (2010) 127–146.

[31] Ambashta R. D., Sillanpää M. Water purification using magnetic assistance: A review, *Journal of Hazardous Materials* 180 (2010) 38–49.
[32] Gómez-Pastora J, Bringas E, Ortiz I. Recent progress and future challenges on the use of high performance magnetic nano-adsorbents in environmental applications, *Chemical Engineering Journal* 256 (2014) 187–204.
[33] Reddy D. H. K., Yun Y-S. Spinel ferrite magnetic adsorbents: Alternative future materials for water purification?, *Coordination Chemistry Reviews* 315 (2016) 90–111.
[34] Trujillo-Reyes J., Peralta-Videa J. R., Gardea-Torresdey J. L. Supported and unsupported nanomaterials for water and soil remediation: Are they a useful solution for worldwide pollution?, *Journal of Hazardous Materials* 280 (2014) 487–503.
[35] Hu J., Lo I. M. C., Chen G. Comparative study of various magnetic nanoparticles for Cr(VI) removal, *Separation and Purification Technology* 56 (2007) 249–256.
[36] Yang L-X., Wang F., Meng Y-F., Tang Q-H., Liu Z-Q. Fabrication and Characterization of Manganese Ferrite Nanospheres as a Magnetic Adsorbent of Chromium, *Journal of Nanomaterials* (2013) 293464.
[37] Zhao X., Wang W., Zhang Y., Wu S., Li F., Ping Liu J. Synthesis and characterization of gadolinium doped cobalt ferrite nanoparticles with enhanced adsorption capability for Congo Red, *Chemical Engineering Journal* 250 (2014a) 164–174.
[38] Muntean C., Stoia M., Barvinschi P. Synthesis of nanocrystalline ZnFe2O4 and its use for the removal of Congo Red from aqueous solutions, *Environmental Engineering and Management Journal* 12 (5) (2013) 959–967.
[39] Afkhami A., Moosavi R. Adsorptive removal of Congo red, a carcinogenic textile dye, from aqueous solutions by maghemite nanoparticles, *Journal of Hazardous Materials* 174 (2010) 98–403.
[40]Chen R, Wang W, Zhao X, Zhang Y, Wu S, Li F, Rapid hydrothermal synthesis of magnetic CoxNi1-xFe2O4 nanoparticles and their application on removal of Congo red, Chemical Engineering Journal 242 (2014b) 226-233.
[41] Liu R., Fu H., Yin H., Wang P., Lu L., Tao Y. A facile sol combustion and calcination process for the preparation of magnetic Ni0.5Zn0.5Fe2O4 nanopowders and their adsorption behaviors of Congo red, *Powder Technology* 274 (2015) 418–425.
[42] Rahimi R., Kerdari H., Rabbani M., Shafiee M. Synthesis, characterization and adsorbing properties of hollow Zn-Fe2O4 nanospheres on removal of Congo red from aqueous solution, *Desalination* 280 (2011) 412–418.
[43] Wang L., Li J., Wang Y., Zhao L., Jiang Q. Adsorption capability for Congo red on nanocrystalline MFe2O4 (M=Mn, Fe, Co, Ni) spinel ferrites, *Chemical Engineering Journal*, 181-182 (2012) 72–79.
[44] Konicki W., Sibera D., Mijowska E., Lendzion-Bielun Z., Narkiewicz U. Equilibrium and kinetic studies on acid dye Acid Red 88 adsorption by magnetic ZnFe2O4 spinel ferrite nanoparticles, *Journal of Colloid and Interface Science* 398 (2013) 152–160.

[45] Liu R., Shen X., Yang X., Wang Q., Yang F. Adsorption characteristics of methyl blue onto magnetic Ni0.5Zn0.5Fe2O4 nanoparticles prepared by the rapid combustion process, *Journal of Nanoparticle Research* 15 (2013) 1679.

[46] Yang L, Zhang Y, Liu X, Jiang X, Zhang Z, Zhang T, Zhang L, The investigation of synergistic and competitive interaction between dye Congo red and methyl blue on magnetic MnFe2O4, Chemical Engineering Journal 246 (2014) 88-96.

[47] Bao X., Qiang Z., Ling W., Chang J. Sonohydrothermal synthesis of MFe2O4 magnetic nanoparticles for adsorptive removal of tetracyclines from water, *Separation and Purification Technology* 117 (2013) 104–110.

[48] Zhang Q., Cheng X., Zheng C., Feng X., Qiu G., Tan W., Liu F. Roles of manganese oxides in degradation of phenol under UV-Vis irradiation: Adsorption, oxidation, and photocatalysis, *Journal of Environmental Sciences* 23 (11) (2011) 1904–1910.

[49] Mamindy-Pajany Y., Hurel C., Marmier N., Romeo M. Arsenic (V) adsorption from aqueous solution onto goethite, hematite, magnetite and zero-valent iron: effects of pH, concentration and reversibility, *Desalination* 281 (1) (2011) 93–99.

[50] Tu Y-J., You C-F., Chang C-K., Wang S-L., Chan T-S. Arsenate adsorption from water using a novel fabricated copper ferrite, *Chemical Engineering Journal* 198-199 (2012) 440–448.

[51] Somorjai, G. and Y. Borodko, Research in nanosciences–great opportunity for catalysis science. Catalysis letters, 2001. 76(1-2): p. 1-5.

[52] Koe, W. S. et al. An overview of photocatalytic degradation: photocatalysts, mechanisms, and development of photocatalytic membrane, *Environmental Science and Pollution Research* (2019) 1–44.

[53] Barick and D. Bahadur. Fe3O4 embedded ZnO nanocomposites for the removal of toxic metal ions, organic dyes and bacterial pathogens, *Journal of Materials Chemistry A* 1 (10) (2013) 3325–3333.

[54] Tju, H., A. Taufik, and R. Saleh. Enhanced UV photocatalytic performance of magnetic Fe3O4/CuO/ZnO/NGP nanocomposites, *Journal of Physics: Conference Series* (2016) IOP Publishing.

[55] Yang, X. et al. Rapid degradation of methylene blue in a novel heterogeneous Fe 3 O 4@ rGO@ TiO 2-catalyzed photo-Fenton system. *Scientific Reports* 5 (2015) 10632.

[56] Mascolo, M.C., Y. Pei, and T. A. Ring. Room temperature co-precipitation synthesis of magnetite nanoparticles in a large pH window with different bases, *Materials* 6 (12) (2013) 5549–5567.

[57] Upadhyay, S., K. Parekh, and B. Pandey. Influence of crystallite size on the magnetic properties of Fe3O4 nanoparticles, *Journal of Alloys and Compounds* 678 (2016) 478–485.

[58] Han, L. et al. A Magnetically Recoverable Fe3O4–NH 2–Pd Sorbent for Capture of Mercury from Coal Derived Fuel Gas, *Scientific Reports* 7 (1) (2017) 1–10.

[59] Bagbi, Y. et al. Lead and chromium adsorption from water using L-cysteine functionalized magnetite (Fe3O4) nanoparticles, *Scientific Reports* 7 (1) (2017) 1–15.
[60] Chen, Y. et al. Magnetically separable Fe3O4@ TiO 2 nanospheres: preparation and photocatalytic activity, *Journal of Materials Science: Materials in Electronics*, 27 (10) (2016) 9983–9988.
[61] Sampaio, M. J. et al. Ag-loaded ZnO materials for photocatalytic water treatment, *Chemical Engineering Journal* 318 (2017) 95–102.
[62] She, P. et al. A self-standing macroporous Au/ZnO/reduced graphene oxide foam for recyclable photocatalysis and photocurrent generation, *Electrochimica Acta* 246 (2017) 35–42.
[63] Feng, Y. et al. Honeycomb-like ZnO mesoporous nanowall arrays modified with Ag nanoparticles for highly efficient photocatalytic activity, *Scientific Reports* 7 (1) (2017) 1–11.
[64] Yang, S., Zhang, M., Wu, X., Wu, X., Zeng, F., Li, Y., Wu, X. The excellent electrochemical performances of ZnMn2O4/Mn2O3: The composite cathode material for potential aqueous zinc ion batteries, *Journal of Electroanalytical Chemistry* (2018). doi:10.1016/j.jelechem.2018.10.05.
[65] R. Schmuch, R. Wagner, G. Horpel, T. Placke, M. Winter. Performance and cost of materials for lithium-based rechargeable automotive batteries, *Nat. Energy* 3 (2018) 267–278.
[66] Q. B. Zhang, H. X. Chen, L. L. Luo, B. T. Zhao, H. Luo, X. Han, J. W. Wang, C. M. Wang, Y. Yang, T. Zhu, M. L. Liu. Harnessing the concurrent reaction dynamics in active Si and Ge to achieve high performance lithium-ion batteries, *Energy Environ. Sci.* 11 (2018) 669–681.
[67] Z. X. He, M. M. Li, Y. H. Li, J. Zhu, Y. Q. Jiang, W. Meng, H. Z. Hou, L. Wang, L. Dai. Flexible electrospun carbon nanofiber embedded with TiO2 as excellent negative electrode for vanadium redox flow battery, *Electrochim. Acta*, 281 (2018) 601–610.
[68] Z. X. He, Y. Q. Jiang, J. Zhu, H. Y. Wang, Y. H. Li, H. Z. Zhou, W. Meng, L. Dai, L. Wang, N-doped carbon coated LiTi2(PO4)3 as superior anode using PANi as carbon and nitrogen bi-sources for aqueous lithium ion battery, *Electrochim. Acta* 279 (2018) 279–288.
[69] K. G. Ramesh, C. I. Juan. A brief review on multivalent intercalation batteries with aqueous electrolytes, *Nanomaterials* 6 (2016) 41.
[70] W. Li, J. R. Dahn, D. S. Wainwright, Rechargeable lithium batteries with aqueous electrolytes, *Science* 264 (1994) 1115–1118.
[71] X. W. Wu, Y. H. Li, C. C. Li, Z. X. He, Y. H. Xiang, L. Z. Xiong, D. Chen, Y. Yu, K. Sun, Z. Q He, P. Chen. The electrochemical performance improvement of LiMn2O4/Zn based on zinc foil as the current collector and thiourea as an electrolyte additive, J. Power Sources 300 (2015) 453–459.
[72] X. Zhu, X. W. Wu, T. N. L. Doan, Y. Tian, H. B. Zhao, P. Chen. Binder-free flexible LiMn2O4/carbon nanotube network as high power cathode for rechargeable hybrid aqueous battery, *J. Power Sources* 326 (2016) 498–504.
[73] N. Zhang, F. Y. Cheng, Y. C. Liu, Q. Zhao, K. X. Lei, C. C. Chen, X. S. Liu, J. Chen. Cation-deficient spinel ZnMn2O4 cathode in Zn(CF3SO3)2

electrolyte for rechargeable aqueous Zn-ion battery, *J. Am. Chem. Soc.* 138 (39) (2016) 12894–12901.

[74] J. J. Cui, X. W. Wu, S. N. Yang, C. C. Li, F. Tang, J. Chen, Y. Chen, Y. H Xiang, X. M. Wu, Z. Q. He. Cryptomelane-type KMn8O16 as potential cathode material-for aqueous zinc ion battery, *Fron. Chem.* (2018). doi: 10.3389/fchem.2018.00352.

[75] X. Guo, G. Z. Fang, W. Y. Zhang, J. Zhou, L. T. Shan, L. B. Wang, C. Wang, T. Q. Lin, Y. Tang, S. Q. Liang. Mechanistic insights of Zn2+ storage in sodium vanadates, *Adv. Energy Mater.* (2018) 1801819. doi: 10.1002/aenm.201801819.

[76] B. Y. Tang, G. C. Fang, J. Zhou, L. B. Wang, Y. P. Lei, C. Wang, T. Q. Lin, Y. Tang, S. Q. Liang. Potassium vanadates with stable structure and fast ion diffusion channel as cathode for rechargeable aqueous zinc-ion batteries, *Nano Energy* 51 (2018) 579 –587.

[77] Y. S. Cai, F. Liu, G. C. Fang, J. Zhou, A.Q. Pan, S.Q. Liang. Pilotaxitic Na1.1V3O7.9 nanoribbons/graphene as high-performance sodium ion battery and aqueous zinc ion battery cathode, *Energy Storage Mater.* 13 (2018) 168–174.

[78] C. Y. Zhu, G. C. Fang, J. H. Guo, Z. Q. Wang, C. Wang, J. Y. Li, Y. Tang, S. Q. Liang. Binder-free stainless steel@Mn3O4nanoflower composite: a high-activity aqueous zinc-ion battery cathode with high-capacity and long-cycle-life, *J. Mater. Chem. A* 6 (2018) 9677–9683.

[79] J. Zhou, L. T. Shan, Z. X. Wu, X. Guo, G. Z. Fang, S. Q. Liang. Investigation of V2O5 as a low-cost rechargeable aqueous zinc ion battery cathode, *Chem. Commun.* 54 (2018) 4457–4460.

[80] C. J. Xu, B. J. Li, H. D. Du, F. Y. Kang, Energetic zinc ion chemistry: the rechargeable zinc ion battery, *Angew. Chem.* 124 (2012) 957–959.

[81] H. L. Pan, Y. Y. Shao, P. F. Yan, Y. W. Cheng, K. S. Han, Z. M. Nie, C. M. Wang, J. H. Yang, X. L. Li, P. Bhattacharya, K. T. Mueller, J. Liu. Reversible aqueous zinc/manganese oxide energy storage from conversion reactions, *Nat. Energy* 1 (2016) 16039.

[82] M. H. Alfaruqi, J. Gim, S. Kim, J. J. Song, J. Jo, S. H. Kim, V. Mathew, J. J. Kim. Enhanced reversible divalent zinc storage in a structurally stable α-MnO2 nanorod electrode, *J. Power Sources* 288 (2015) 320–327.

[83] B. Lee, C. S. Yoon, H. R. Lee, K. Y. Chung, B. W. Cho, S. H. Oh. Electrochemically-induced reversible transition from the tunneled to layered polymorphs of manganese dioxide, *Sci. Rep.* 4 (2014) 6066–6074.

[84] B. Lee, H. R. Lee, H. Kim, K. Y. Chung, B. W. Cho, S. H. Oh. Elucidating the intercalation mechanism of zinc ions into α-MnO2 for rechargeable zinc batteries, *Chem. Commun.* 51 (2015) 9265–9268.

[85] Q. Wenda, Y. Li, A. You, Z. Zhang, G. Li, X. Lu and Y. Tong. High-performance flexible quasi-solid-state Zn-MnO2 battery based on MnO2 nanorod arrays coated 3D porous nitrogen-doped carbon cloth, *J. Mater. Chem. A* 5 (2017) 14838–14846.

[86] H. F. Li, G. P. Han, Y. Huang, M. S. Zhu, Z. X. Pei, Q. Xue, Z. F. Wang, Z. X. Liu, Z. J. Tang, Y. K. Wang, F. Y. Kang, B. H. Li, C. Y. Zhi. An extremely

safe and wearable solid-state zinc ion battery based on a hierarchical structured polymer electrolyte, *Energy Environ. Sci.* 11 (2018) 941–951.

[87] M. Y. Yan, P. He, Y. Chen, S. Y. Wang, Q. L. Wei, K. N. Zhao, X. Xu, Q. Y. An, Y. Shuang, Y. Y. Shao, K. T. Mueller, L. Q. Mai, J. Liu, J. H. Yang. Water-lubricated intercalation in V2O5·nH2O for high-capacity and high-rate aqueous rechargeable zinc batteries, *Adv. Mater.* 30 (2018) 1703725.

[88] X. W. Wu, Y. H. Xiang, Q. J. Peng, X. S. Wu, Y. H. Li, R .C. Song, F. Tang, Z. X. Liu, Z. Q. He, X. M. Wu. A green-low-cost rechargeable aqueous zinc-ion battery using hollow porous spinel ZnMn2O4 as the cathode material, *J. Mater. Chem. A* 5 (2017) 17990–17997.

[89] S. D. Han, S. Kim, D. G. Li, V. Petkov, H. D. Yoo, P. J. Phillips, H. Wang, K. L. More, B. Key, R. F. Klie, J. Cabana, V. R. Stamenkovic, T. T. Fister, N. M. Markovic, A. K. Burrell, S. Tepavcevic, J. T. Vaughey,#. Mechanism of Zn insertion into nanostructured δ- MnO2: A aonaqueous rechargeable Zn metal battery, *Chem. Mater.* 29 (2017) 4874–4884.

[90] M. H. Alfaruqi, S. Islam, D. Y. Putro, V. Mathew, S. Kim, J. Jo, S. Kim, Y. K. Sun, J. Kim. Structural transformation and electrochemical study of layered MnO2 in rechargeable aqueous zinc-ion battery, *Electrochim. Acta* 276 (2018) 1–11.

[91] B. Z. Jiang, C. J. Xu, C. L. Wu, L. B. Dong, J. Li, F. Y. Kang. Manganese sesquioxide as cathode material for multivalent zinc ion battery with high capacity and long cycle life, *Electrochim. Acta* 229 (2017) 422–428.

[92] Doaga, A., Cojocariu, A. M., Amin, W., Heib, F., Bender, P., Hempelmann, R., and Caltun, O. F. Synthesis and characterizations of manganese ferrites for hyperthermia applications. *Materials Chemistry and Physics* 143 (1) (2013) 305–310. doi:10.1016/j.matchemphys.2013.08.

[93] Shah, S. A., Majeed, A., Shafique, M. A., Rashid, K., and Awan, S.-U. Cell viability study of thermo-responsive core-shell superparamagnetic nanoparticles for multimodal cancer therapy, *Applied Nanoscience* 4 (2) (2013) 227–232. doi:10.1007/s13204-012-0191-8.

[94] P. T. Phong, N. X. Phuc, P. H. Nam, N. V Chien, D. D. Dung, P. H. Linh. Condensed Matter Size-controlled heating ability of CoFe2O4 nanoparticles for hyperthermia applications, *Phys. B Phys. Condens. Matter.* 531 (2018) 30–34. doi:10.1016/j.physb.2017.12.010.

[95] S. Ammar, A. Helfen, N. Jouini, F. Fie, Ë. Villain, P. Molinie, M. Danot, Â. P. Diderot, P. Jussieu, F. E-mail, Â. P. Marie, P. Jussieu, I. J. Rouxel, C. de Houssinie. *Magnetic properties of ultra ® ne cobalt ferrite particles synthesized by hydrolysis in a polyol medium* (2001) 186–192. doi:10.1039/b003193n.

[96] A. Cristina, S. Samia. Structural effects on the magnetic hypertermia properties of iron oxide nanoparticles, *Prog. Nat. Sci. Mater. Int.* 26 (2016) 440–448. doi:10.1016/j.pnsc.2016.09.004.